国家自然科学基金资助项目（4181101168，51874268）
江西理工大学清江学术文库

Estimating and Predicting Pore-pressure Influence on Deep-seated Landslides

Wen Nie　Kui Zhao

Beijing
Metallurgical Industry Press
2019

Copyright© 2019 by Metallurgical Industry Press, China

Published and distributed by Metallurgical Industry Press
39 Songzhuyuan North Alley, Beiheyan St Beijing 100009, P. R. China

All rights reserved. No part of this publication may be reproduced, stored in a retrieval system, or transmitted in any form or by any means, electronic, mechanical, photocopying, recording or otherwise, without the prior written permission of the copyright owner.

图书在版编目（CIP）数据

孔隙水压力对大型滑坡的影响预测＝Estimating and Predicting Pore-pressure Influence on Deep-seated Landslides：英文/聂闻，赵奎著．—北京：冶金工业出版社，2019.7
ISBN 978-7-5024-8142-1

Ⅰ.①孔… Ⅱ.①聂… ②赵… Ⅲ.①孔隙水压力—影响—滑坡—研究—英文 Ⅳ.①P642.22

中国版本图书馆 CIP 数据核字（2019）第 110176 号

出 版 人　谭学余
地　　址　北京市东城区嵩祝院北巷 39 号　邮编　100009　电话　(010)64027926
网　　址　www.cnmip.com.cn　电子信箱　yjcbs@cnmip.com.cn
责任编辑　徐银河　美术编辑　彭子赫　版式设计　孙跃红
责任校对　郑　娟　责任印制　李玉山

ISBN 978-7-5024-8142-1
冶金工业出版社出版发行；各地新华书店经销；北京建宏印刷有限公司印刷
2019 年 7 月第 1 版，2019 年 7 月第 1 次印刷
169mm×239mm；9.25 印张；161 千字；138 页
59.00 元

冶金工业出版社　投稿电话　(010)64027932　投稿信箱　tougao@cnmip.com.cn
冶金工业出版社营销中心　电话　(010)64044283　传真　(010)64027893
冶金工业出版社天猫旗舰店　yjgycbs.tmall.com

(本书如有印装质量问题，本社营销中心负责退换)

Foreword

Groundwater-pressure produced by precipitation plays an important role in determining landslide movements. For any process-related early warning system for hydrologically-induced landslides, the accurate estimation of groundwater-pressure from precipitation information and the related movement is a key component. Until now, the problem of how to predict the pore-pressure and following movement by the precipitation has no satisfactory solution, because of the complex nonlinear hydrogeological and geological system. But using advanced conceptual models and machine learning algorithm, the author of this book has carried out some cases applications to fill the knowledge gap. The book is an integration of research results over the past five years mainly based on the doctor book. The project obtained support from alpEWAS project (im Programm Geotechnologien Federal Ministry of Education and Research, Germany) and Opening Fund of State Key Laboratory of Geohazard Prevention and Geo-environment Protection (Chengdu University of Technology)-SKLGP2013K007. Some of the experimental and theoretical results have been published in major international journals, such as Natural Hazards and Earth System Sciences, Bulletin of Engineering Geology and the Environment, and Geofluids. The international peer review process for these English language journals is an additional guarantee of quality.

All things considered, I believe that the publication of this book offers a sound modelling basis for developments in prediction of hydrologically-triggered deep-seated landslides. It increases our understanding of the application of machine learning algorithm and uncertainty model in hydrologically-triggered deep-seated landslide movement. I am pleased to commend this book to all interested readers.

Meifeng Cai

Feb. 2018

Preface

This book aims to construct a landslide forecasting component for precipitation data to predict groundwater-pressure and related landslide movement.

First, a description of how to predict precipitation types is introduced including snow accumulation/snow melt models. Hydro-models are discussed for the evaluation of infiltration and pore water pressure produced by ground water supply. Second, a modified tank model for the estimation of the groundwater table is developed which is i) coupled with a snow accumulation/snow melt model, ii) and accounts for equivalent infiltration, a time lag elimination method, to reduce uncertainty about precipitation and infiltration. The application of this modified tank model is demonstrated for the Aggenalm landslide in Bavaria, Germany. Third, physical model experiments evaluate the work modes of three typical tank models for homogenous materials. Fourth, in terms of heterogeneous materials, the physical tank model experiments consider four typical geological conditions. Next, an estimation of groundwater-related quasi-staticlandslides is developed. It is often assumed that groundwater level controls the increase and decrease of this landslide movement velocity and that it reacts according to Coulomb friction and is velocity independent. In this book, an approach is presented, where the strength of the slip surface and especially its cohesion is considered a velocity-dependent factor. A new viscous model is constructed by introducing a velocity dependent cohesion to describe the landslide movement rate controlled by groundwater changes. Two cases are used to validate the new model, those of the Undercliff on the Isle of Wight in England and the Utiku landslide in New Zealand. This book provides important steps to in future better understand and model the effect of groundwater level and pore pressure changes on landslides movement. This could significantly contribute to future model-based rather than present threshold-based early warning systems.

Finally, I also offer my special thanks to Prof. Michael Krautblatter for his valuable and constructive suggestions, especially during the stage of application of modified tank model. My sincere thanks go to my mentor Dr. Kerry Leith. He helped me review almost every detail of quasi-static landslide movement model and kept discussing with me once per week at least. And the same thanks to monitoring data about Underclif, Isle of Wight Landslide from Dr. Jonathan Martin Carey (GNS Science, New Zealand), Prof. David Petley (University of East Anglia, UK).

Wen Nie

Jan. 2019.

Contents

1 **Introduction** 1
   1.1 Problem statement and motivation 1
   1.2 Research objectives 1
   1.3 Book outline 2
   1.4 Technical course 3

2 **Literature review** 7
   2.1 Overview 7
   2.2 Classification of landslides 7
   2.3 Causes of landslides 8
   2.4 Hydrological controls of deep-seated landslides 9
   2.5 Hydrogeological flow patterns relevant for deep-seated landslides 10
   2.6 Estimating proportions of fluid and solid precipitation and snowmelt 11
      2.6.1 Thresholds for fluid and solid precipitation 11
      2.6.2 Temperature-index and energy-index snowmelt models 12
   2.7 Estimating infiltration and groundwater fluctuations in hillslopes 13
      2.7.1 Deterministic physical models 13
      2.7.2 Empirical-statistical models 14
   2.8 Modelling quasi-static landslide movements (slope movement with complete sliding surface) 17

3 **A modified tank model including snowmelt and infiltration time lags for deep-seated landslides** 21

3.1	Introduction	22
3.2	Site descriptions	24
	3.2.1 Geographical setting and geological overview	24
	3.2.2 Tectonic overview and setting of the study site	26
	3.2.3 Climatic conditions	26
	3.2.4 Monitoring system and monitoring data	28
	3.2.5 Historic events	30
3.3	Methods	34
	3.3.1 The modified tank model including snowmelt and infiltration	34
	3.3.2 Simpler approximations of slope hydrology	39
	3.3.3 Determining the parameter of PWP calculation in the modified tank model	39
	3.3.4 Snowmelt calculations in the modified tank model	44
3.4	Results	45
	3.4.1 Performance of modified tank model in heavy rainfall season	45
	3.4.2 Performance of modified tank model in snowmelt season	45
	3.4.3 Performance of modified tank model throughout the monitoring period and error analysis	46
3.5	Discussions	48
	3.5.1 Performance of modified tank model in heavy rainfall season	48
	3.5.2 Performance of modified tank model in snowmelt season	48
	3.5.3 Highlights of the modified model	49
	3.5.4 Drawbacks and limitations	49
3.6	Conclusions	50

4 Physical tank experiments on groundwater level controls of slopes with homogenous materials 51

4.1	Introduction	51
4.2	Methods	52
	4.2.1 Test setup and testing materials	52
	4.2.2 Experiment procedures	59
4.3	Results and analysis	60

4.3.1		Simple tank experiment	60
4.3.2		Surface runoff tank experiment	61
4.3.3		Lateral water flow supply tank experiment	63
4.4	Conclusions	64

5 Physical tank experiments for estimation of groundwater considering slope structure controlling affection ... 65

- 5.1 Introduction ... 65
- 5.2 Different typical geological condition of landslides ... 65
  - 5.2.1 A coarse-fine material slope ... 65
  - 5.2.2 A slope including a fine layer ... 65
  - 5.2.3 A slope including an obvious fracture ... 68
  - 5.2.4 A slope in interaction with a river ... 68
- 5.3 Physical tank experiments ... 70
  - 5.3.1 Tank experimental flume and materials ... 71
  - 5.3.2 Physical tank experimental outline ... 76
- 5.4 Results and discussion ... 76
- 5.5 Summary and conclusions ... 81

6 Prediction of groundwater affecting deep-seated landslide quasi-static movement ... 82

- 6.1 Introduction ... 82
- 6.2 New viscous velocity based model ... 85
  - 6.2.1 Introduction of new viscous model ... 85
  - 6.2.2 Velocity-strength module ... 88
  - 6.2.3 New viscous model calculation ... 90
- 6.3 The Ventnor landslide, Isle of Wight, Southern England ... 91
  - 6.3.1 Introduction to Ventnor landslide, Isle of Wight ... 91
  - 6.3.2 Monitoring data in the model (displacement and pore water pressure) ... 91
  - 6.3.3 Strength properties of the main slip surface ... 93
  - 6.3.4 Application of a new viscous model at the Ventnor landslide ... 94

  6.3.5 Prediction results and analysis ··················· 95
6.4 The Utiku landslide in New Zealand ················ 96
  6.4.1 Introduction to Utiku landslide, New Zealand ············ 96
  6.4.2 Involved monitoring and strength date ················ 98
  6.4.3 Application of new viscous model to Utiku landslide ········ 98
  6.4.4 Prediction results and analysis ··················· 101
6.5 Discussion ······························· 101
  6.5.1 Physical interpretation ······················· 101
  6.5.2 Parameters interpretation ····················· 104
  6.5.3 Comparison between traditional and new viscous model ······ 105
  6.5.4 Error analysis and outlooks ···················· 110
6.6 Conclusions ····························· 110

7 Conclusion ································ 111
7.1 Key findings ····························· 111
7.2 Limitations and outlook ······················· 112

References ································· 113

Appendix I Codes of landslide model ················ 126
 1 Cohesion back analysis in quasi-static landslide model ·········· 126
 2 Quasi-static landslide prediction (cohesion constant) ··········· 128
 3 Quasi-static landslide prediction (cohesion-velocity simple
   linear) ································· 131

Appendix II List of symbols ····················· 135

1 Introduction

1.1 Problem statement and motivation

With the expansion of modern infrastructure and societal development in mountainous areas, landslide hazards have becomea serious issue in all mountain environments worldwide. Hydrological triggering is by far the most frequent landslide triggering mechanism, and related hazards cause major damage and more than 1000 casualties every year (Giannecchini et al., 2012; Guzzetti et al., 2007; Guzzetti et al., 2008). It is recognised that pore water pressure (PWP) changes related to fluctuating groundwater tables play a critical role in controlling deep-seated landslide instability, because a rise in PWP causes a decrease in effective normal stress at the potential sliding surface (Bromhead, 1978; Bonnard and Noverraz, 2001; Wang and Sassa, 2003; Collins and Znidarcic, 2004; Schulz et al., 2009; Rahardjo et al., 2010). Pore water pressure measurements produced by groundwater table fluctuations have traditionally been used to explain and sometimes even predict landslide velocities (e.g. Corominas et al., 2005). The effect of groundwater, and especially long-term groundwater conditions, can prepare and trigger deep-seated landslides (Burda and Vilímek, 2010). Although the processes linking rainfall, snow melt, slope hydrology, and deep-seated landslide movements are still difficult to establish, the development of an accurate and straightforward prediction of pore pressure changes and a corresponding landslide movement model is of great importance for any landslide forecasting framework (Malet et al., 2005).

1.2 Research objectives

In order to develop a forecasting framework of hydrologically-induced (hereafter abbreviated as hydro-induced) landslides to issue a reliable prediction, it is necessary to construct an accurate and simple groundwater prediction model that is capable of incorporating snow accumulation, snow melt and infiltration time lags. The next step would be a displacement prediction model controlled by fluctuating groundwater tables. Combining those two, a forecasting system for hydro-induced landslides should be able to predict landslide acceleration and deceleration controlled by pore pressures changes resulting from rainfall and snow precipitation. As an important component of an enhanced landslide forecasting system, the aim and scope of this book is to develop an accurate and simple groundwater prediction model and a groundwater-landslide movement prediction model.

To achieve a model like this, the following steps were taken:

(1) The data from rainfall gauge monitoring and a snow accumulation/snow melt model based on temperature-index was used for precipitation calculation.

(2) Relationships between precipitation, infiltration, previous water content, and groundwater in time domains were analysed.

(3) A simple and effective modified tank model was developed for the prediction of pore pressure changes based on historical monitoring data.

(4) Relationships between the displacement rate, groundwater, and material strength in a quasi-static landslide analysis were established and modelled.

(5) Changing strength-velocity relations were incorporated in the model to better describe the landslide movement rate under variable groundwater conditions and in response to dynamic material weakening.

1.3 Book outline

Following this introductory chapter, the book comprises seven chapters arranged accordingly to the objectives mentioned in Section 1.2.

In Chapter 2, a comprehensive literature review introduces the state of the art of landslide classifications, precipitation-related landslides, precipitation calculation models, groundwater calculation models, and deep-seated quasi-static stability analysis.

In Chapter 3, a modified tank model for the estimation of groundwater is put forward. This model considers the time lag between precipitation and groundwater changes and includes snow accumulation/snowmelt and the creation of an infiltration path. An equivalent infiltration calculation method is employed to reduce time lag error. By considering water flow supply and drainage, the modified tank model can predict the groundwater caused by single or a series of precipitation events. A slow deep-seated landslide at Aggenalm, in Bavaria, Germany is chosen for applying and testing the model.

Chapter 4 deals with groundwater investigations in physical tank experiments under simulated rainfall using homogeneous soils, considering three tank modes. Rainfall infiltration with or without surface runoff is first investigated. Lateral water flow supply coupled with infiltration affecting the groundwater is then considered. These experiments emphasise the causes and phenomena of infiltration time lags.

Chapter 5 describes groundwater investigations in physical tank experiments involving four types of complex geological structures.

In Chapter 6, a new viscous landslide model is constructed for the prediction of landslide quasi-static movement. Compared to the original viscous landslide model, this viscous model has following advantages:

(1) Using a limit equilibrium model rather than the infinite slope model for force analysis means that the applications of models are wider.

(2) Slope density is a function of the groundwater table, not a constant, which makes the driving forces of slopes variable.

(3) Material strength is designed to be dependent on velocity reflecting the soil consolidation and shear dilation action. Two landslide cases, in the UK and in New Zealand, are used to validate the new landslide model.

In Chapter 7, a synbook of the research is presented. The methods developed throughout the book and the results obtained from experiments and modelling are summarised and evaluated. Limitations of the methods are presented, together with suggestions for future model improvements and research needs.

1.4 Technical course

The book involves three main contents:

(1) Realisation of a conceptual hydro-caused landslide forecasting system.

(2) Estimation or prediction of groundwater-pressure by a modified tank model (groundwater model).

(3) Prediction of quasi-static landslide movement velocity of hydro-induced landslides using a new viscous (movement-rate) landslide model.

Figure 1.1 provides an overall overview of the model components in the book. The model indicating near future changes in pore pressure is explained in detail in Figure 1.2. The model indicating near future landslide displacement rates is drafted in Figure 1.3.

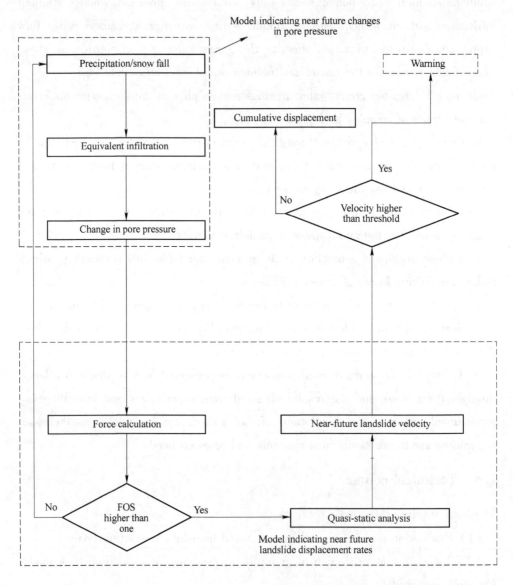

Figure 1.1 Realization of a conceptual hydro-induced landslide forecasting system. It includes the modelling near future changes in pore water pressure and modelling near future landslide displacement rates

1.4 Technical course

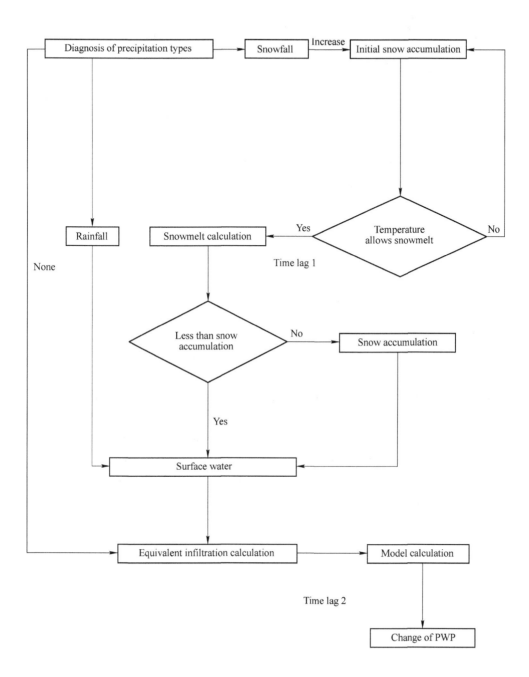

Figure 1.2 Modified tank model for the estimation of near future changes in pore pressure. It includes diagnosis of precipitation types, calculation of snow accumulation and melt, equivalent infiltration calculation, and calculation of change of PWP in slip surface

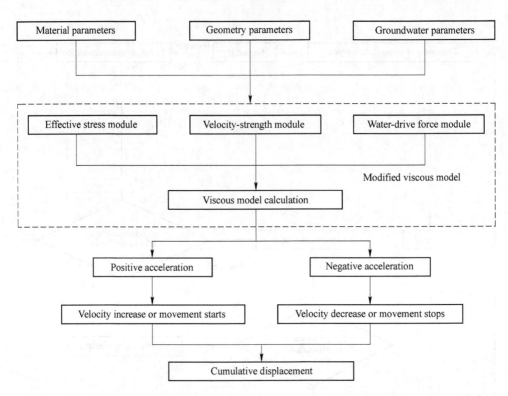

Figure 1.3 Model indicating near future landslide displacement rates derived from the modified viscous landslide model for hydro-induced quasi-static landslide movements. The input data for the model calculation includes initial material, geometry, and groundwater parameters; the model has an effective stress module affected by groundwater-pressure, a velocity-strength module (considering the velocity affecting the material strength), and a water-drive force module (the landslide drive force varies with the density of slope mass, which is caused by change of groundwater)

2 Literature review

2.1 Overview

The study presented in this book relates to hydrological, geological and geotechnical sciences, modelling technology, and optimisation theory. Relevant issues of these subjects are hereafter briefly reviewed. The first part of this chapter gives a brief description of landslide classifications. Common empirical, conceptual and instability mechanism of precipitation-related deep-seated landslides are then reviewed, followed by basic concepts of hillslope hydrology and hydrogeology. The second part of the literature review covers the current understanding of snow accumulation and melt models. Infiltration and groundwater calculation models, including deterministic models and probability models are introduced. The third part of the literature review emphasises the process of the quasi-static problem of deep-seated landslides. Three models are demonstrated for quasi-static landslide analysis: the impulse response model, phenomenological relationship model, and viscous-plastic velocity-based model.

2.2 Classification of landslides

Knowledge of landslide classification is important to avoid misunderstanding, and to structure the different processes (Cruden and Varnes, 1996). Variable landslide characteristics produce numerous landslide classifications for different purposes. The most commonly established landslide classifications are based on criteria such as type of failure, morphology, movement, activity and material.

Cruden (1991) and Cruden and Varnes (1996) defined a landslide as "a mass of rock, debris or earth sliding down a slope". Hutchinson (1988) focused on the morphology of slope movements, considering the mechanisms, materials and rates of the movement. Cruden (1991) and Cruden and Varnes (1996) classified landslides mainly according to their movement types (fall, slide, flow, spread), material and the shape of the failure plane. Hungr et al. (2001) modified this classification for flow-type landslides, taking soil saturation into account. Lateltin et al. (1997) proposed to categorise landslides according to the depth of the slip surface: 0~2m below surface: shallow landslides; 2~10m below surface: medium seated landslides; and more than 10m below surface: deep-seated landslides.

Dikau (1996) provided a simple division into seven distinct types (Table 2.1).

Table 2.1 Mass movement classification (Dikau, 1996)

Type	Rock	Debris	Soil
Fall	Rock fall	Debris fall	Soil fall
Topple	Rock topple	Debris topple	Soil topple
Slide (rotational)	Single (sump) multiple successive	Single multiple successive	Single multiple successive
Slide (translational)	Block slide	Block slide	Slab slide
Planar	Rock slide	Debris slide	Mudslide
Lateral spreading	Rock spreading	Debris spread	Soil (debris) spreading
Flow	Rock flow (Sackung)	Debris flow	Soil flow
Complex (with run-out or change of behaviour downslope)	Rock avalanche	Flow slide	Slump earth flow

A "fall" is defined as a free-fall movement of material from a steep slope or cliff. "Topples" involve a pivoting action rather than a complete separation at the failure base. "Sliding movements" involve material displacement along one or more discrete shear surfaces, and are subdivided into "rotational" or "translational slides" according to the geometry of the failure surface (Dikau, 1996). "Rotational" landslides occur along a circular slip surface, whereas "translational" failures have a planar slip surface, often bedding controlled. "Lateral spreading" involves low angled slopes and often display slow movement rates. A "flow" failure can occur over a spread of slope angles and customarily behaves similarly to fluids, due to the significant volumes of water or air involved. "Complex" landslide failures, in this classification, are failures comprising a combination of different failure mechanisms (Table 2.1).

2.3 Causes of landslides

When considering individual failure events, landslide failure can be understood in terms of forces. The actual causes of landslides are complex, due to the diversity of factors acting upon and within a slope. These causes may be subdivided into two groups (Table 2.2): internal causes leading to a reduction in shear strength, and external causes producing an increase in the shear stress.

Table 2.2 **Summary of causes of landslide failures** (after Moore et al., 2006)

Internal Causes	External Causes
Materials:	Removal of slope support:
1. Soils subject to strength loss on contact with water or as a result of stress relief (strain softening)	1. Undercutting by water (waves and stream incision)
2. Fine grained soils which are subject to strength loss or gain due to weathering	2. Washing out of soil (groundwater flows)
3. Soils with discontinuities characterised by low shear strength such as bedding planes, faults, joints etc.	3. Human intervention (e.g. cutting, excavations, tunnelling and mining)
Weathering:	Increased loading:
1. Physical and chemical weathering of soils causing loss of strength (apparent cohesion and friction)	1. Natural accumulations of water, snow, talus
2. Slope ripening and soil weakening processes (e.g. loss of vegetation, shrink and swell, desiccation and surface cracking)	2. Human intervention (e.g. fill, tips, buildings)
Pore-water pressure:	Transient Effects:
Elevated pore-water pressures causing a reduction in effective shear stress. Such effects are most severe during wet periods or intense rainstorms	1. Earthquakes and tremors
	2. Shocks and vibrations

2.4 Hydrological controls of deep-seated landslides

Precipitation, subsequent infiltration and groundwater flow are some of the most important landslide triggering factors (Johnson and Sitar, 1990; Leroueil et al., 1996; Noverraz and Bonnard, 1998; Van Asch et al., 1999). Usually, a rising groundwater table within a saturated zone, which is frequently observed during heavy rainfall, leads to a gradual increase of pore water pressure in the soil (Iverson, 1997). An increase in the pore water pressure decreases the effective normal stresses in the soil and the total stress remains constant under drained conditions. This reduces the resistant strength and destabilises the slope. The lag time depends on the permeability of the soil, hydraulic conductivity, depth of slope, and previous moisture. It has also been shown that in clayey soil, pore water pressure changes rapidly due to increased permeability induced by the increase of cracks, fissures, and lenses of more permeable material (Rogers and Selby, 1980; Duncan et al., 2014). Increasing positive water pressure on the slip surface from an increasing groundwater level can cause deeper landslides (5~20m deep) (Van Asch et al., 1999).

2.5 Hydrogeological flow patterns relevant for deep-seated landslides

The main aim of hillslope hydrology is to study the flow paths and lag time of rainwater from a catchment to a receiving stream or an infiltration area. Water from fluid (rainfall) and solid precipitation (snowmelt) either flows as surface runoff or infiltrates the soil. Figure 2.1 shows different flow paths of water on and in a hillslope.

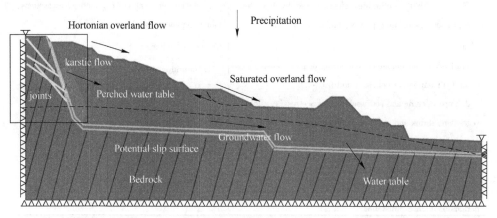

Figure 2.1 Schematic diagram of a hillslope hydrological system. It includes hortonian overland flow, saturated overland flow, perched water table, karstic flow, and groundwater flow

Two types of surface runoff are distinguished: Hortonian overland flow occurs when precipitation exceeds the infiltration rate. Saturated overland flow occurs when the soil has reached complete saturation. Bedrock is often assumed to be an impermeable boundary below the sediment (Beven and Germann, 1982; Brammer and McDonnell, 1996), which is questionable especially for fractured and bedrock affected by karst phenomena. Groundwater flow infiltrates and percolates through saturated underground soil along the bedrock (Weiler et al., 2005). Gravity mainly drives the water flow (McDonnell et al., 2007). Groundwater flow is more obvious in saturated or nearly saturated soil because of the connectivity in macro pores rather than micro pores in unsaturated soil (Beven and Germann, 1982). Many important aquifers are composed of granular materials such as loose sand and gravel or weakly cemented bedrock (Figure 2.1). Groundwater flow in these aquifers is through the pores or spaces between the grains of sand or gravel, or through narrow fractures in solid bedrock. How quickly the water flows is partly dependent on how big the pores are, how interconnected the pores or fractures are, and how much energy (head or water pressure) is available to move the water through the aquifer. Generally, granular aquifers or fractured bedrock aquifers have no equivalent "underground river" or channel. In a karst aquifers, the fractures (secondary porosity) are formed by joints, bedding planes, and faults (see the red square in Figure 2.1). The openings forming the karst aquifers may be partly or completely water-filled. Karstic flow in limestones probably dominates the subsurface flow regime in reality.

2.6 Estimating proportions of fluid and solid precipitation and snowmelt

2.6.1 Thresholds for fluid and solid precipitation

A threshold temperature, under which precipitation can be considered solid, is a key factor in any kind of snow accumulation model. However, this temperature is difficult to achieve, because the vertical travel height of snow flakes and the temperature of snow flakes cannot easily be calculated nor acquired. It has long been recognised that the vertical temperature profile is of prime importance for the type of precipitation (see, for example, Wagner, 1957; Bocchieri, 1980; Czys et al., 1996). In some situations, a temperature variation of only 1℃ is sufficient to cause a transition between different phases, i.e. snow and rain. This implies that an accurate vertical temperature profile forecast is needed to correctly determine precipitation types. Precipitation begins as snow if the echo top is in air colder than 0℃, otherwise it starts as rain. A wet-bulb temperature, which is always lower than the air temperature in an unsaturated atmosphere, is used instead of a dry-bulb, as this more accurately estimates the temperature at the surface of a hydrometeor. If the wet-bulb temperature is great than 1.3℃, the precipitate melts to rain (Albers et al., 1996). The technique proposed by Ramer (1993) uses temperature, relative humidity, and wet-bulb temperature on different pressure levels to diagnose precipitation types. This procedure performs two checks before making a full calculation. If the surface wet-bulb temperature is greater than 2℃, liquid precipitation is diagnosed. If wet-bulb temperature remains below a specified value at all levels, solid precipitation is expected. A full calculation is needed if neither condition is satisfied. In fact, in some areas, snowstorms often begin with a surface air temperature near 2℃ (36℉). For falling snowflakes to survive in air with temperatures significantly above the freezing point, the air must be unsaturated (relative humidity less than 100%), and the wet-bulb temperature must be at the freezing point (Ahrens, 2007). Although temperature is most important factor, any ideal diagnosis for the precipitation type will take all related atmospheric parameters into account. One way to include all the physics associated with the precipitation type is to develop a sophisticated and explicit microphysics model. However, these models are expensive in computer time, thus limiting their operational applications and in the meantime other alternatives must be used to routinely forecast precipitation types. The most common approach is still to derive statistical relationships between some predictors, such as temperature, and different precipitation types (Bourgouin, 2000). To validate such approaches, the rainfall information is usually collected by a rainfall gauge directly, if the diagnosis of precipitation shows rainfall not snowfall. If the diagnosis of precipitation indicates snowfall from the beginning, a rain gauge with a heat device can melt the snowfall, e. g. in winter. Solid precipitation is thus estimated in the form of liquid precipitation for the calculation of snow heights, considering the temperature and humidity.

2.6.2 Temperature-index and energy-index snowmelt models

Snow melt models can usually be grouped into two main categories.

The most sophisticated models are spatially distributed models based on equations of mass and energy balance (Bloschl et al., 1991; Garen and Marks, 2005; Herrero et al., 2009). These models, following a mechanistic approach, account for multiple physical and chemical processes in the snowpack. Such models are quite complex and require physical parameters, including topography, precipitation, air temperature, wind speed and direction, humidity, incoming shortwave and longwave radiation, cloud cover, and surface pressure. The determination of accurate values of these parameters, and their variation in space and time is only possible for very well-equipped experimental test sites (Lakhankar et al., 2013).

Simplified approaches, such as temperature-index methods, are also widely used

(Kustas et al., 1994; Rango and Martinec, 1995; Hock, 1999, 2003; Jost et al., 2012). These models use air temperature as an index to perform an empirical correlation with snowmelt, and require only a few parameters (e. g. precipitation, air temperature, snow covered area). Temperature-index achieves good results and some additional improvement in model performance when coupling an energy balance approach (Hock, 2003). One of the most popular methods used to forecast snowmelt is to correlate air temperature with snowmelt data. Such a relationship was first used for an Alpine glacier by Finsterwalder (1887) and has since then been widely applied and further refined (Herrmann, 1978; Kustas et al., 1994; Rango and Martinec, 1995; Hock, 1999, 2003; Jost et al., 2012).

2.7 Estimating infiltration and groundwater fluctuations in hillslopes

Generally, there are two ways to estimate groundwater changes:

(1) Deterministic models depending on the permeability and infiltration of material, such as the Green and Ampt model (Chen and Young, 2006), and the Richards equation coupled with Van Genuchten equation (Van Genuchten, 1980; Schaap and Van Genuchten, 2006; Weill et al., 2009), which need detailed parameterisation considering material and geological structure.

(2) Empirical-statistical models using optimisation or fitting parameters such as the Tank model or the HBV (Hydrologiska Byråns Vattenbalansavdelning) model, which needs historical monitoring data for training parameters (Faris and Fathani, 2013; Abebe et al., 2010).

2.7.1 Deterministic physical models

Green and Ampt (1911) presented an approach based on fundamental physics which is capable of calculating the wetting front in the soil materials in accordance with empirical observations. This model can e. g. be used to determine how much water was added to the soil with time.

Richards (1931) equation represents the movement of water in unsaturated soils. Richards applied a continuity requirement and obtained a general partial differential equation describing water movement in unsaturated non-swelling soils. It is a non-linear partial differential equation, which is often difficult to approximate since it does not have a closed-form analytical solution.

The shape of water retention curves in the Richards equation can be characterised using several models, one of the most prominent is the Van Genuchten Model (Schaap and Van Genuchten, 2006; Weill et al., 2009). The water retention curve is the relationship between the water content and the soil water potential. This curve is characteristic of different types of soil, and is also called the soil moisture characteristic. It is used to predict the soil water storage, water supply to plants (field capacity) and soil aggregate stability. Due to the hysteretic effect of water filling and draining of the pores, different wetting and drying curves may be distinguished. At potentials close to zero, a soil is close to saturation, and water is held in the soil primarily by capillary forces. As moisture decreases, the binding of the water becomes stronger, and at small potentials (more negative, approaching wilting point) water is strongly bound in the smallest of pores, at contact points between grains, and as films bound by adsorptive forces around particles. Sandy soils will mainly involve capillary binding, and will therefore release most of the water at higher potentials, while clayey soils, with adhesive and osmotic binding, will release water at lower (more negative) potentials. At any given potential, peaty soils will usually display much higher moisture contents than clayey soils, which would be expected to hold more water than sandy soils. Thus, the water holding capacity of any soil is due to the porosity and the nature of the bonding in the soil.

2.7.2 Empirical-statistical models

The linear reservoir concept is based on analysis of the recession limbs of the drainage hydrographs, and has already been used extensively for description of catchment responses (Hornberger et al., 1991; Dingman, 1994; Sivapalan et al., 2002). The basic linear storage theory can be described as:

$$Q = kS \qquad (2\text{-}1)$$

$$dS/dt = -Q \qquad (2\text{-}2)$$

where, Q is the outflow; S is the storage amount of water; t is the time and k is the rate constant. T indicates the buffering capacity of a reservoir or the "slowness" of water release.

Similarly, for the groundwater table estimation, the HBV model and the tank model based on the linear storage theory are introduced here as follows:

The HBV model (Bergström, 1995; Bergström et al., 2001) is a rainfall-runoff model which includes conceptual descriptions of hydrological processes at a catchment scale. The model is normally run on daily values of rainfall and air temperature, and daily or monthly estimates of potential evaporation. The model consists of subroutines for meteorological interpolation, snow accumulation and melt, evapotranspiration estimation, soil moisture accounting procedures, routines for runoff generation, and finally, a simple routing procedure between sub-basins and lakes. The soil routine in the HBV model is depicted by the soil zone and two reservoirs. The upper reservoir is responsible for the fast runoff at the intensive precipitation, while the lower one dominates in times without precipitation.

The widely used tank model is a complex linear theorised calculation to describe the behaviours of water hydraulic properties (Ishihara and Kobatake, 1979). It is based on the water balance theory that tracks water into and out of a particular area of interest (Figure 2.2). The basic tank model usually can simulate the groundwater of one point in a shallow slope without considering the lateral water flow supply.

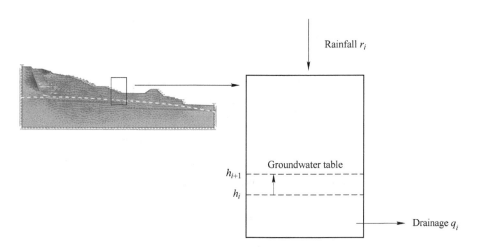

Figure 2.2 Schematic diagram of simple tank model. It considers the relationship among rainfall infiltration, drainage, and change of groundwater table in one time unit

$$h_{i+1} - h_i = r_i - q_i \quad (2\text{-}3)$$

$$q_i = ah_i \quad (2\text{-}4)$$

where, h_{i+1} and h_i are the ground water tables of $i+1$th day and ith day; r_i and q_i are the rainfall and drainage of ith day; a is a relation parameter.

In this book, tank models are applied for the groundwater table estimation in a deep-seated landslide. Due to the long infiltration path in deep-seated landslides, the rainfall (r_i) cannot entirely contribute to the change of groundwater table ($h_{i+1} - h_i$) within one day. A long time lag can be expected to affect the prediction of groundwater table.

Considering the lateral groundwater flow, infiltration time lag, and surface runoff, until recently, a multi-storage tank model was developed to estimate the groundwater fluctuations of landslides caused by heavy rainfall (Ohtsu et al., 2003; Takahashi, 2004; Takahashi et al., 2008; Xiong et al., 2009) (Figare. 2. 3). In Figure 2. 3, the left tank is used for the simulation of the higher part of a slope, while the right tank can simulate the lower part of a slope.

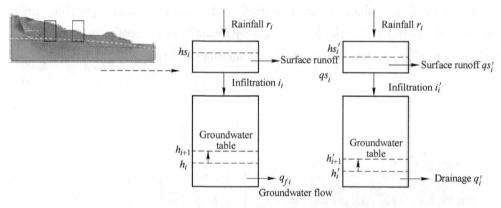

Figure 2. 3 Schematic diagram of multi-storage tank model. The multi-storage tank model mainly simulates a complex hydro-system which involves surface runoff, infiltration, groundwater flow, drainage, and groundwater table

$$h_{i+1} - h_i = i_i - q_{fi} \tag{2-5}$$

$$q_{fi} = ah_i \tag{2-6}$$

$$i_i = r_i - qs_i \tag{2-7}$$

$$qs_i = bhs_i \tag{2-8}$$

$$h'_{i+1} - h'_i = q_{fi} + i'_i - q'_i \tag{2-9}$$

$$q'_i = ch'_i \tag{2-10}$$

$$i'_i = r_i - qs'_i \qquad (2\text{-}11)$$

$$qs'_i = dhs'_i \qquad (2\text{-}12)$$

where, h_{i+1} and h_i are the upper ground water table of $i+1$th day and ith day; r_i and q_{fi} are the rainfall and groundwater flow of ith day; i_i is the upper infiltration of ith day; qs_i is the upper surface runoff of ith day; hs_i is the upper surface water of ith day; h'_{i+1} and h'_i are the lower ground water table of $i+1$th day and ith day; r_i and q'_i are the rainfall and drainage of ith day; i'_i is the lower infiltration of ith day; qs'_i is the lower surface runoff of ith day; hs'_i is the lower surface water of ith day; a, b, c, d are relation parameters.

A problem of multi-storage tank models is the presence of unknown parameters. Even the usage of some advanced algorithms does not effectively train the parameters. An increase of unknown parameters would reduce the robustness and reliability of the system, especially if single validation datasets are used, such as monitored groundwater table for the parameter estimation of whole system. Thus, the innovation of this approach (In Chapter 3) is to calculate the equivalent infiltration before it enters the tank. The equivalent infiltration deals with the infiltration time lag including snow accumulation and snowmelt in deep-seated landslides based on a simple tank model structure. It hypothesizes that, compared to a simple tank model, the modified model has a higher accuracy and physical meaning by controlling equivalent infiltration including snow accumulation and snowmelt. Compared to multi-tank models the modified model is more robust and reliable.

2.8 Modelling quasi-static landslide movements (slope movement with complete sliding surface)

Landslide movements usually evolve over very longer times, and landslides would undergo active, stop, and reactive phases (Auzet and Ambroise, 1996; Swanston et al., 1995; Vlcko et al., 2009; Wienhofer et al., 2009). Fast movements, acceleration, and final failure could then be triggered e.g. by extreme rainfall or snowmelt. According to Angeli et al. (2004) and Picarelli (2007), there are three chronological stages for a deep-seated landslide before final failure: the processes of static instability, development of whole slip surface, and quasi-static landslide movement. As shown in Figure 2.4, in Time Step 1 the slope has a local instability (brittle damage) induced by a change of the water table. As time elapses (Time Step 2 and Time Step 3), the whole potential slip surface is developed. Then the slope gets into the quasi-static landslide stage which means that the landslide produces a displacement not a local or tiny deformation. In Time Step 4, the landslide moves very slow due to a lower groundwater table. In Time Step 5, the landslide would accelerate due to an increase of groundwater table.

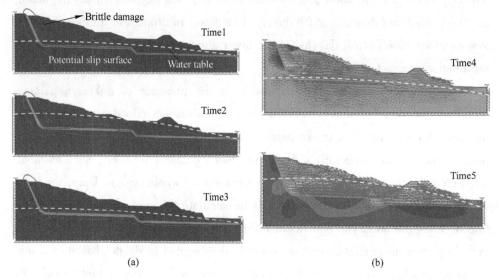

Figure 2.4 Schematic diagram of deep-seated landslide failure

(a) Static instability stage. The whole slip surface develops from a local brittle damage to a global slip surface;

(b) Quasi-static instability stage. The landslide controlled by resisting and drive forces moves as a whole

For the estimation of quasi-static landslides, the following three models are often used: The impulse response model (IR model) is based on the use of a black box model originally dedicated to hydro geological and hydro geochemical data analysis (Pinault and Schomburgk, 2006). It permits the processing of data and model temporal series, with computation of transfer functions between input data and output data, based on signal process methods, inversion and optimisation techniques.

Vulliet and Hutter (1988) proposed a phenomenological relationship in which the base sliding velocity is a function of the ratio between the shear stress acting along the failure plane, and the shear strength, is in turn a function of the normal effective stress.

Viscous models are widely used in quasi-static landslides because they can calculate acceleration or deceleration movement under the groundwater table change, which is more akin to real situations (Figure 2.5) (Angeli et al., 1996; Angeli et al., 1998; Angeli et al., 1999; Leroueil et al., 1996; Corominas et al., 2005; Herrera et al., 2009; Di Maio et al., 2013).

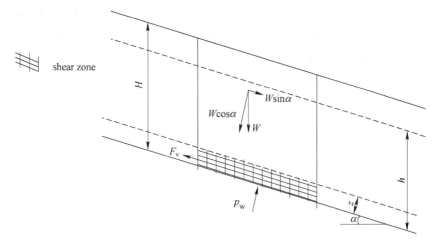

Figure 2.5 Geometry and variables used in the viscous landslide model. The model is based on the infinite slope model and can calculate a local displacement rate in one time unit

α—slope angle; H—hight of slope mass; h—water table; W—weight;

F_v—viscous force; z—shear zone thickness; p_w—pore water pressure

The original viscous model can be expressed by Equation(2-12) and Equation(2-13).

$$F - T' = ma_c + F_v = ma_c + \eta \frac{v}{z} \qquad (2\text{-}13)$$

$$\gamma H \sin\alpha \cos\alpha - [c + (\gamma H \cos^2\alpha - p_w)\tan\phi] = ma + \eta \frac{v}{z} \qquad (2\text{-}14)$$

where, F and T' are slip force and resistance force of slope unit; m is the mass of the shear zone; a_c is the acceleration; F_v is viscous force; η is the viscosity; v is the velocity; z in our case is the height of the shear zone; γ is the unit weight of soil mass; H is the depth of soil layer to the potential failure surface; α is the slip surface inclination; c, ϕ, p_w represents the effective soil cohesion, the effective friction angle, and pore water pressure at the potential slip surface. In this model, c and γ are always treated as constants.

The second main object of this book is to improve the prediction of quasi-static deep-seated landslide movements based on viscous landslide models. Traditionally, in the original viscous model, changes of normal stress caused by groundwater fluctuation and precipitation infiltration are regarded as a main factor to control the movement velocity. In Chapter 6, the driving forces exerted by water and especially the velocity-dependent cohesion are regarded for landslide movements, which can reduce the cumulative prediction error of the viscous-landslide model.

3 A modified tank model including snowmelt and infiltration time lags for deep-seated landslides

Deep-seated landslides are an important and widespread natural hazard within alpine regions, and can have significant impacts on infrastructure. Pore water pressure plays an important role in determining the stability of hydrologically triggered deep-seated landslides. Based on a simple tank model structure, the groundwater level prediction is improved by introducing time lags associated with groundwater supply caused by snow accumulation, snowmelt, and infiltration in deep-seated landslides. In this study, an equivalent infiltration calculation is demonstrated to improve the estimation of time lags using a modified tank model to calculate regional groundwater levels. Applied to the deep-seated Aggenalm Landslide in the German Alps at 1000 ~ 1200m, the results predict daily changes in pore water pressure ranging from -1kPa to 1.6kPa depending on daily rainfall and snowmelt which are validated by piezometric measurements in boreholes. The inclusion of time lags improves the results of standard tank models by around 36% (linear correlation with measurement) after heavy rainfall and, respectively, by around 82% following snowmelt in a 1 ~ 2 day period. For the modified tank model, this chapter introduces a representation of snow accumulation and snowmelt, based on a temperature index and an equivalent infiltration method, i.e. the melted snow-water equivalent. The modified tank model compares well to borehole-derived water pressures. Changes of pore water pressure can be modelled with 0% ~ 8% relative error in rainfall season (standard tank model: 2% ~ 16% relative error) and with 0% ~ 7% relative error in snowmelt season (standard tank model: 2% ~ 45% relative error). Here, a modified tank model is demonstrated for deep-seated landslides which includes snow accumulation, snow melt and infiltration effects and can effectively predict changes in pore water pressure in alpine environments.

3.1 Introduction

Deep-seated landslides in the European Alps and other mountain environments pose a significant hazard to people and infrastructure (Mayer et al., 2002; Madritsch and Millen, 2007; Agliardi et al., 2009). It has long been recognized that pore water pressure (PWP) changes due to precipitation play a critical role for hydrologically controlled deep-seated landslide activation. The rise in PWP causes a drop of effective normal stress on potential sliding surfaces (Bromhead, 1978; Iverson, 2000; Wang and Sassa, 2003; Rahardjo et al., 2010). The estimation of pore water pressure is of great significance for anticipating deep-seated landslide stability. In past years, geotechnical monitoring systems have revealed PWP changes related to rainfall and snowmelt events (Angeli et al., 1988; Simoni et al., 2004; Hong et al., 2005; Rahardjo et al., 2008; Huang et al., 2012). Generally, two ways are employed to estimate the groundwater changes:

(1) Depending on the precise information of permeability and infiltration of material, the Green and Ampt Model is generally used to describe groundwater infiltration and water table changes (producing PWP) in saturated homogeneous material (Chen and Young, 2006); the Richards Equation (Weill et al., 2009) with the Van Genuchten Equation (Van Genuchten, 1980; Schaap and Van Genuchten, 2006) and the Fredlund and Xing (1994) method show better performance in the evaluation of infiltration and groundwater table in unsaturated homogeneous material. Traditional deterministic models have advantages due to their explicit physical and mechanical approaches, but they require accurate knowledge, testing and monitoring of soil physical parameters which are often not available with sufficient accuracy. For example, the widely used Richards Equation with the Fredlund and Xing method needs soil suction tests under variable moisture content which is difficult to achieve for complex landslides with multiple reworked materials.

(2) Empirical-statistical models employ optimization or fitting parameters in their model structure. Tank and other models need historical monitoring data to train parameters (Faris and Fathani, 2013; Abebe et al., 2010). Such probabilistic models, because of their simple conceptualized structure, do rely to a smaller degree on explicit physical and mechanical approaches. However, they can avoid the prob-

lems induced by the uncertainty of material parameterisation and their spatial arrangement in the landslide mass. They can, therefore, be applied to a wide range of different landslide settings and we estimate that for more than 90% of all landslides no explicit parameters on soil suction etc. are available. As one of the most common probabilistic models, tank models typically describe infiltration and evaporation in shallow soil materials (Ishihara and Kobatake, 1979). They are based on the water balance theory, which means they account for flows into and out of a particular drainage area. Multi-tank models involving two or three tank elements have been developed to better estimate groundwater fluctuations within shallow landslides induced by heavy rainfall (Michiue, 1985; Ohtsu et al., 2003; Takahashi, 2004; Takahashi et al., 2008; Xiong et al., 2009).

Simple tank models do not consider infiltration time lags induced by a long infiltration path, previous moisture and snowmelt. This inhibits their applicability to deep-seated landslide. Multi-tank models can deal with infiltration time lags to some extent by adding tanks but even then they require data from several monitoring boreholes to track groundwater flow supplies in complicated geological structures and they are presently not designed to replicate time lags of increased infiltration, e. g., following snowmelt (Iverson, 2000; Sidle, 2006; Nishii and Matsuoka, 2010). Applying multi-tank models to compensate for time lags is questionable as especially deep-seated landslides would need several tanks to replicate time lags and every added new tank in vertical direction introduces 3 new parameters at least. This would reduce robustness and reliability of the system especially if just the monitored groundwater table is used for the parameter training of the whole system. In this study, a simple method is introduced to estimate time lags by a modified standard tank model which predicts changes in pore water pressure. The innovation of this approach is to calculate equivalent infiltration before it enters the tank. The equivalent infiltration deals with the infiltration time lag including snow accumulation and snowmelt in deep-seated landslides based on a simple tank model structure. It hypothesizes that, compared to a simple tank model, the modified model has a higher accuracy and physical meaning by controlling equivalent infiltration including snow accumulation and snowmelt; compared to multi-tank model our modified model is more robust and reliable. The modified tank model is applied to the Aggenalm landslide, where predicted

PWP changes can be tested against piezometric borehole monitoring data. The monitoring network design and installation, as well as detailed monitoring data, and the introduction of monitoring devices have been described previously in detail (Thuro et al., 2009; Thuro et al., 2011a; Thuro et al., 2011b; Festl et al., 2011; Thuro et al., 2013).

3.2 Site descriptions

3.2.1 Geographical setting and geological overview

The Aggenalm Landslide is situated in the Bavarian Alps of the Aggenalm Region near Bayrischzell. Bayrischzell is located in the south of Bavaria, about 80km southeast of Munich and 30km south of Rosenheim at the northern edge of the Alps within the Mangfall mountain range (Figure 3.1). The tectonic map of Aggenalm region is as shown in Figure 3.2. The study site is an element of the mountainous Aggenalm region between the towns of Bayrischzell and Oberaudorf and also two major valleys, the Ursprung valley in the west and the Auerbach valley in the east. The undulating Aggenalm area is framed by the southern Brünnstein-Traithen ridge and the northern Wildbarren-Wendelstein ridge. The Aggenalm landslide has a maximal length of 780m and the maximum width of 340m. The elevation of the main scarp of the landslide is about 1200m and at its toe about 920m.

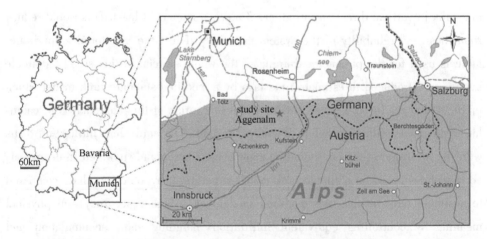

Figure 3.1 Location of the study site at the Aggenalm landslide in the Bavarian Alps of the Aggenalm Region near Bayrischzell, Germany (Festl, 2014)

3.2 Site descriptions

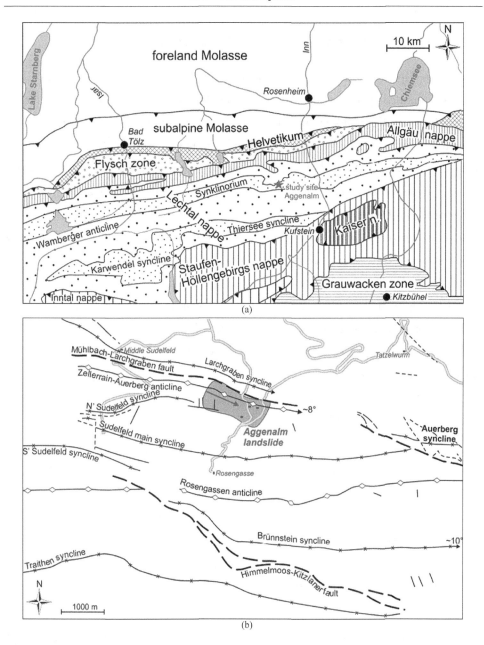

Figure 3.2 The tectonic map of Aggenalm region
(a) Tectonic map of the Northern Calcareous Alps between Lake Starnberg and Chiemsee. The Aggenalm Landslide is situated in the Lechtal Nappe within the Synklinorium, a major syncline-anticline-syncline fold belt, which can be traced through the whole region (Schmidt-thome, 1964; Gwinner, 1971); (b) Detailed tectonic map showing the main tectonic features in the Aggenalm landslide area. Here the Synklinorium has a complex structure with several additional minor syn- and anticlines, of which the eastward dipping of the Zellerrain-Auerberg Anticline is responsible for the nearly slope parallel orientation of the rock mass within the Sudelfeld landslide (Festl, 2014)

3.2.2 Tectonic overview and setting of the study site

During the Alpine orogeny, the rock mass was faulted and folded into several large east-west oriented synclines, of which the Audorfer Syclinorium is responsible for the nearly slope-parallel bedding orientation of the rock mass in the area of the Aggenalm Landslide (Figure 3.3). The Aggenalm Landslide is underlain by Late Triassic well-bedded limestones (Plattenkalk, predominantly Nor), overlain by Kössen Layers (Rhät, predominantly marly basin facies) and the often more massive Oberrhät Limestones and Dolomites (Rhät) (Figure 3.3). The marls of the Kössen Layers are assumed to provide primary sliding surfaces and are very sensitive to weathering as they decompose over time to a clay-rich residual mass (Nickmann et al., 2006). The landslide mechanism can be classified as a complex landslide dominated by deep-seated sliding with earth flow and lateral rock spreading components (Singer et al., 2009). A major activation of the landslide occurred in 1935, destroying three bridges and a local road. Slow slope deformation and secondary debris flow activity has been ongoing since this time.

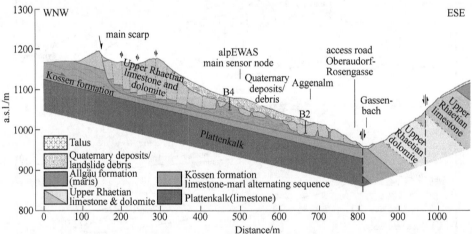

Figure 3.3 Geological profile of the Aggenalm Landslide based on the data and new information acquired, e.g., by the drilling campaign and geoelectric survey (modified from Festl, 2014)

3.2.3 Climatic conditions

The Aggenalm is exposed to a subcontinental climate with a pronounced summer precipitation maximum and an annually changing share of 15%~40% of the mean annual precipitation that fall as snow. Nearby meteo-stations such at the Brünnsteinhaus, the Sudelfeld (Polizeiheim) and the Tatzelwurm show mean annual precipitation of 1594mm/y, 1523mm/y and 1660mm/y at similar elevations (Table 3.1).

Table 3.1 **Mean annual precipitation** (1931~1960 and 1961~1990) **at different observation stations** (Wolff, 1985)

Observation station	Mean annual precipitation/mm · y^{-1}
Degerndorf-Brannenburg	1346
Oberaudorf	1398
Bayrischzell	1403
Brünnsteinhaus	1594
Sudelfeld (Polizeiheim)	1523
Tatzelwurm	1660
Wendelstein	1814

The data sets from the Brünnsteinhaus station (1594mm/y) as well as the Sudelfeld (Polizeiheim) station (1523mm/y) and Tatzelwurm station (1660mm/y) can represent best the conditions at the Aggenalm landslide because all three observation stations are at a similar elevation and at similarly oriented slopes. Table 3.2 indicates the monthly distribution of the precipitation (rainfall/snowfall) for the above stations.

Table 3.2 **Mean monthly precipitation** (1931~1960 and 1961~1990) **for the Brünnsteinhaus, the Sudelfeld** (Polizeiheim), **and Tatzelwurm observation stations** (data from Germany's National Meteorological Service)

Location	Precipitation	Oct.	Nov.	Dec.	Jan.	Feb.	Mar.	Winter
Brünnsteinhaus(1345m)	mm	89.6	109.2	115.7	102.9	100.5	103.2	621.1
Sudelfeld (Polizeiheim) (1070m)	mm	84.7	98.3	113.1	82.7	82.2	95.5	556.5
Tatzelwurm(795m)	mm	109.9	106.1	99.1	123.2	118.7	110.9	667.9
Location	Precipitation	Apr.	May.	June.	July.	Aug.	Sep.	Summer
Brünnsteinhaus(1345m)	mm	121.9	138.4	194.3	208.6	193.0	116.8	973.0
Sudelfeld (Polizeiheim) (1070m)	mm	103.7	152.3	204.2	195.1	199.2	112.0	966.5
Tatzelwurm(795m)	mm	115.8	149.4	194.8	224.9	185.6	121.9	992.4

3.2.4 Monitoring system and monitoring data

The main three monitoring systems include time domain reflectometry (TDR) to monitor shear displacements of subsurface, reflectorless video tacheometry (VTPS) and a low-cost global navigation satellite system (GNSS) to detect displacements on the surface of the landslide. Well-established techniques-including inclinometers, crack meters, and extensometers for deformation measurements as well as two piezometers and a weather station, including a rain gauge to monitor precipitation were installed at the Aggenalm landslide. Figure 3.4 shows the location of the main measuring devices installed on surface and in boreholes at the Aggenalm landslide as well as the most important infrastructure elements of the alpEWAS geosensor network. This study focuses on the weather information and pore water pressure. Therefore, especially the two types of monitoring data are emphasized on in the section below.

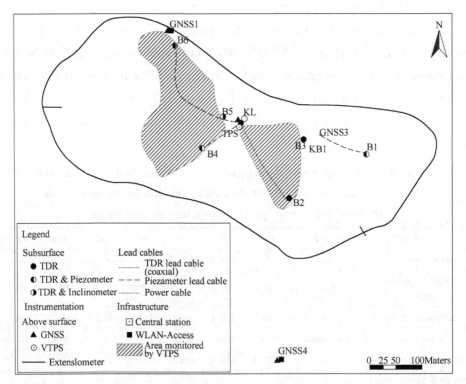

Figure 3.4 Orthophoto of the deep-seated Aggenalm landslide, outlined red. The various measuring devices, and the geosensor network's infrastructure elements, and the area in the viewshed of the reflectorless video tacheometry are marked on the image (Thuro et al., 2010)

The weather station has recorded data since summer 2008 and only a few times data losses occurred (mainly before the start of the automatic data acquisition). Monitoring data for this study was derived from a rain gauge and humidity sensor (alpEWAS central station), and a pore water pressure (PWP) sensor was installed in boreholes close to the assumed shear zone (B4, 29.4 meter deep) (Unfortunately, pore water pressure of B6 was only recorded for several months because of the device broken) (Figure 3.3) (Singer et al., 2009; and Festl, 2014). A heated precipitation gauge provided data on the snow-water equivalent of snowfall. Short term noise in raw data was filtered. PWP, temperature, and humidity were averaged over a 24-hour period (Festl, 2014). Since the whole monitoring period lasted for almost 3 years and time lags are in the range of days, days were considered to be the most robust and appropriate time unit. The monitoring period lasted almost from Feb. 2009 to Dec. 2011 (Figure 3.5). Considering data loss in some months, there are approximately 24 months

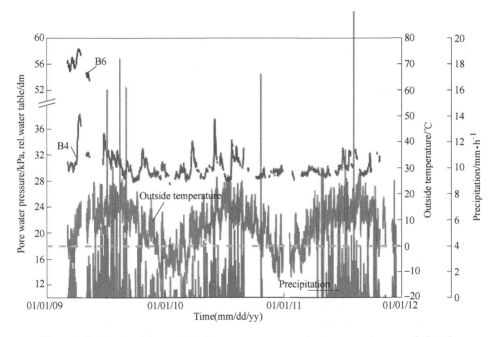

Figure 3.5 Graph of the precipitation, temperature, and piezometer data recorded at the Aggenalm landslide from 2009 through 2011. The pink dashed line marks 0℃ for easier interpretation. All precipitation is considered as snowfall if the temperature is below 0℃ (the data is from alpEWAS project https://www.alpewas.bgu.tum.de/home)

of valid data. The data from the 13 months (May 2009 to June 2009; September 2009 to December 2009; February 2010 to August 2010) was used to parametrize the modified tank model. To validate the parametrized model, 55 days of rainfall (July 2009 to August 2009) and 44 days of snowmelt (March 2009 to April 2009) were used to compare model-calculated pore water pressure with real pore water pressure readings. In addition, simulation of two years PWP levels were compared to the whole two years of monitoring data of PWP levels bridging the data gaps.

The temperature data was used to approximately discern between snowfall and rainfall because both were measured using the rain gauge (with a heater to melt the snow) (Figure 3.5). In 2011 the heater for the rain gauge, had not been installed yet, thus leading to a major data gap. The barometric pressure measurements were necessary to adjust the piezometer readings to the barometric pressure on site and thus to calculate correct water table heights.

3.2.5 Historic events

Only two greater magnitude events happened in the past: 1935 and 1997 (Figure 3.6). Bernrieder (1991) mentions damages in 1899 to the paths and bridges in the area, while Marklseder (1935) refers to recurrent movements at different time intervals.

3.2.5.1 The 1935 event——the Aggenalm landslide

The greatest documented event happened on April 22, 1935, in the eastern Aggenalm area. Major parts of the Aggenalm landslide acted and accelerated producing the main scarp. Waller (n.d.) presented a graphic depicting of the approximate dimensions of the mass movement at the time (incorporated into Figure 3.6 as a light blue dotted line). The event of 1935 was also described by Malalse (1951) and Avo (n.d.). The event's main scarp located about 90m above the Aggenalpe in an area of moraine-covered Kössen formation as well as Lower Jurassic marls (Liasfleckenmergel). In Avo (n.d.), almost two million cubic meters slope mass, weighing five million tons with a maximum depth of 30m, moved along an approximately 250m long sliding surface toward the Gassenbach. Parts of the masses were moved by the creek (Gassenbach) up to 700m to the confluence with the Auerbach. Triggering factor of the

3.2 Site descriptions

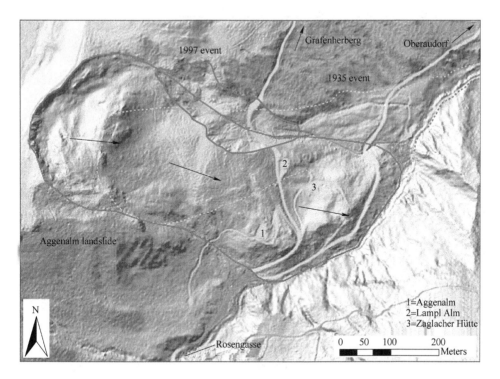

Figure 3.6 Hillshade model of the Aggenalm landslide based on a digital elevation model (DGM, 1m laser scan). The deep-seated mass movement (red line) is monitored by the alpEWAS system and shows only very slow movement rates of 1~2cm per year, expressed by the red arrows. The dimensions of the two documented events from 1935 and 1997 are sketched in blue colors. The location of the three alpine lodges is shown as well (Geobasisdaten © Bayerische Vermessungsverwaltung 2010; schematic movement vectors adapted from Gallemann (2012); outline of 1935 event adapted from Waller (n. d.))

landslide was snowmelt or a combination of snowmelt and heavy rainfall (Marklseder, 1935; Malalse, 1951; Bernrieder, 1991). Figure 3.7 shows that no rainfall prior to the first signs of the mass movement (April 22, 1935) was recorded for Bayrischzell, and Tatzelwurm. However, heavy rainfall was recorded in the weeks before slope failure. In total, 192% of the 30-year average of rainfall was recorded at the Tatzelwurm station in April (Figure 3.7). This above-average rainfall combined with the snowmelt (snow fall in the months prior e. g. , in February 207%) could be causative for the 1935 landslide event. There is a time lag between the beginning of landslide movement and heavy precipitation. It is assumed that the time lags due to the heavy precipitation infiltration.

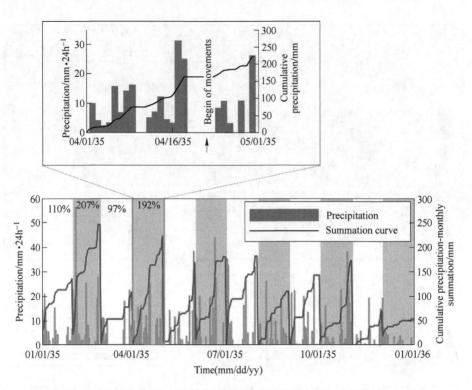

Figure 3.7 The lower graph shows the daily rainfall (light blue) and the monthly summation curve (blue) at the Tatzelwurm observation station in 1935. The percentages printed across the first four months of the year represent the amount of rainfall in comparison to the 30-year average from 1931~1960 (from Festl, 2014)

3.2.5.2 The 1997 event——the Agggraben debris flow

The smaller second event happened in 1997. The Water Management office (Wwa) of Rosenheim states that during fall of 1997 a period of heavy rainfall triggered about 30.000m³ soil/debris and marls originating from the northern edge of the Aggenalm (Wwa rosenheim n. d.). The approximate extent of this debris flow is depicted by a blue line in Figure 3.8.

The annual rainfall of 1997 (1579mm) is slightly above the 30-year mean. Most months are drier than on average, but in July about twice and in October 1.75 times as much rainfall than usual. Since the exact date of the event is not known the cause of the debris flow might be a precedent high water table with a subsequent strong rainfall event (in July, August, or October), triggering the debris flow.

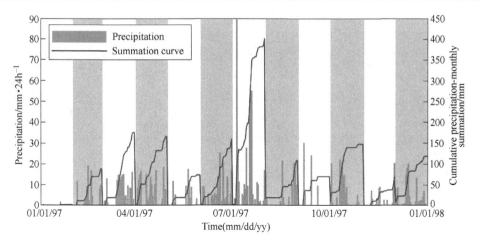

Figure 3.8 The amount of precipitation at the Sudelfeld (Polizeiheim) weather station for the year 1997 is depicted in light blue. The dark-blue lines show the monthly summation curves of rainfall (from Festl 2014)

3.2.5.3 Other landslide events

Parts of the tarmac of the connection road to the Rosengasse were destroyed several times due to small rotational slides during the last years. Small debris flows originating from the Aggenalm landslide mass or the Agggraben debris flow could be observed. The Agggraben debris flow happened in June 2013 after intense rainfall (320mm within 3 days) as shown in Figure 3.9.

(a)

Figure 3.9 The Agggraben debris flow happened in June 2013 after intense rainfall

(a) Small debris flow at the southern edge of the Aggenalm landslide. The debris flow has a width of approx. 3~4m and occurred between late fall 2010 and spring 2011; (b) Translational slide at the connection road to Oberaudorf in consequence of the heavy rainfall at the beginning of June 2013. Width: approx. 20m, length: 15~20m, depth: max. 2m; (c) ~ (e) Debris flow occurring on June 2, 2013, in the Agggraben at the northern edge of the Aggenalm landslide. Its scarp is located within the main scarp from 1997; (c) The material was transported in the space of the bed of the old Agggraben debris flow; (d) and ran over and accumulated on the connection road to Grafenherberg; (e) (Photos (c~e) by courtesy of Stefan Schuhbäck) (from Festl, 2014)

3.3 Methods

3.3.1 The modified tank model including snowmelt and infiltration

Figure 3.10 demonstrates the changes from the original tank model (Ishihara and Kobatake, 1979; Michiue, 1985; Ohtsu et al., 2003; Uchimura et al., 2010) to the modified model. Figure 3.10 (a) shows the basic concept of the original tank model,

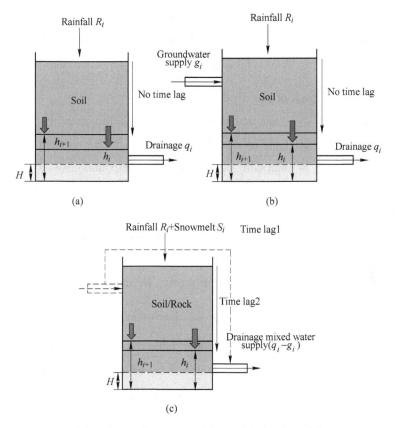

Figure 3.10 Generation of the modified tank model
(a) Original tank model considering only the vertical infiltration and drainage affecting the water table;
(b) Considering both vertical infiltration and horizontal ground water flow; (c) Modified tank model including water supply and two time lags (snowmelt and infiltration)

the daily change in the groundwater table height $h_{i+1} - h_i$ is

$$h_{i+1} - h_i = R_i - q_i \tag{3-1}$$

where, R_i is the rainfall and q_i is the drainage of the ith day; h_i is groundwater table height the ith day.

Concepts illustrated in Fig. 3.10 (b) are now incorporated in the water flow supply tank model including groundwater supply. The daily change in groundwater table height $h_{i+1} - h_i$ is

$$h_{i+1} - h_i = R_i - (q_i - g_i) \tag{3-2}$$

where, g_i is groundwater supply of the ith day from the upper slope.

Another aspect, snowmelt also plays an important role in producing groundwater supply in Figure. 3.10 (c). Thus, the Equation (3-3) should be written as

$$h_{i+1} - h_i = R_i + S_i - (q_i - g_i) \qquad (3\text{-}3)$$

where, S_i is the snowmelt of the ith day.

More importantly, snow accumulation and snowmelt produces the first time lag (time lag 1) as a result of the effects of ambient temperature on snowmelt. Both the groundwater response to snowmelt and rainfall are compounded by long infiltration paths as, for example, water infiltration often takes one or more days to reach the water table in deep-seated landslide masses (time lag 2) (Figure 3.10 (c)). This can be described: the infiltration in ith day does not only affect the groundwater table in ith day but also the groundwater table over the following n days if the time lag 2 (n days) is considered ($n>1$). In other words, R_i and S_i are divided into n parts ($R_i = \sum_{n=1}^{N} R_i^{(n)}$ and $S_i = \sum_{n=1}^{N} S_i^{(n)}$, $i, n \geq 1$). Each component ($R_i^{(n)}$ and $S_i^{(n)}$) contributes to daily changes in the groundwater table ($h_{i+n} - h_{i+n-1}$). Thus, total daily variations ($h_{i+2} - h_{i+1}$) in response to rainfall and snowmelt can be described by $R_{i-1}^{(3)} + S_{i-1}^{(3)}$, $R_i^{(2)} + S_i^{(2)}$, $R_{i+1}^{(1)} + S_{i+1}^{(1)}$ when time lag is 2 days as shown in Figure 3.11, con-

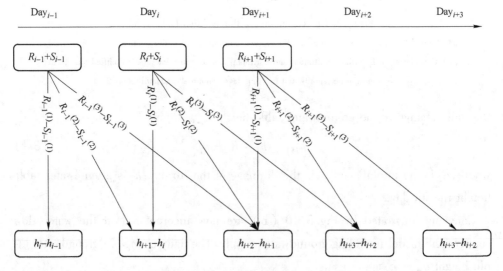

Figure 3.11 Schematic diagram of water infiltration from the surface to the groundwater table producing time lag 2 (time lag is 2 days). Groundwater changes on a given day (Day_{i+1}) result from infiltration over the previous three days (Day_{i-1}, Day_i, and Day_{i+1})

sidering that the groundwater table in $i+1$th day is not only affected by the infiltration today but also the infiltration of previous two days.

The Antecedent Precipitation Index (API) can reduce this time lag 2 by estimating the current water content of the ground affected by previous precipitation (Chow, 1964). The current water content of the ground including rainfall and snowmelt will be treated as the total infiltration of the current day. This is equivalent to the infiltration calculations of Suzuki and Kobashi (1981); Matsuura et al. (2003); Matsuura et al. (2008) who define equivalent infiltration as

$$ER_i + ES_i = (0.5)^{1/M} R_i + (0.5)^{1/M} ER_{i-1} + (0.5)^{1/M} S_i + (0.5)^{1/M} ES_{i-1} \tag{3-4}$$

where, ER_{i-1} and ES_{i-1} represent the equivalent rainfall and snowmelt of $i-1$th days, respectively; R_i and S_i mean the rainfall and snowmelt of ith day; $(0.5)^{1/M}$ means the effect of infiltration reduces to 50% in M days (where M is determined from field observations). Therefore, the whole modified tank model with an equivalent infiltration method could substitute both, time lag 1 by integrating snow accumulation and snowmelt and time lag 2. The introduction of equivalent rainfall and snowmelt for dealing with time lag is much more convenient than by the model calibration for solving the time lag problems.

The relationship between infiltration and water table is often proportional in slopes (Matsuura et al., 2008; Schulz et al., 2009; Thuro et al., 2010; Yin et al., 2010). Therefore, the conceptual equation of changed water table should be like:

$$\Delta h_i = h_{i+1} - h_i = \frac{\alpha}{n}(ER_i + ES_i) - (q_i - g_i) \tag{3-5}$$

where, α is a proportional coefficient (only for ideal tank model, α is one), n is the average porosity of the slope mass. It has to be emphasized on the "pore water pressure" is positive pressure induced mainly by groundwater table height. It is neither perched water table nor negative pore water pressure in unsaturated layer.

Thus, PWP can be linearly correlated to groundwater levels such that:

$$\Delta PWP_i = \frac{\alpha g'}{n}(ER_i + ES_i) - \Delta PWP_{(g+q)i} \tag{3-6}$$

where, g' is acceleration of gravity and $\Delta PWP_{(g+q)i}$ is the PWP changed by subsurface inflows and outflows on the ith day. This allows us to evaluate changes in PWP resulting from infiltration, drainage, and groundwater supply. The major part of pore water pressure is static pressure induced by water table height. Minor components are seepage force and the difference of pressures in the available pore space over drier and wetter periods. Since the tank model is a "grey box model". We do not know the exact proportions of static pressure, seepage pressure, and pressure dynamics in pore space, which are all three included in the equivalent pore water pressure.

$$\Delta PWP_i = \alpha'(ER_i + ES_i) - \Delta PWP_{(g+q)i} \qquad (3\text{-}7)$$

In Equation (3-7) α' replaces $\dfrac{\alpha g'}{n}$ to simplify the model. The workflow chart of our modified tank model for change of PWP_i is indicated in Figure 3.12.

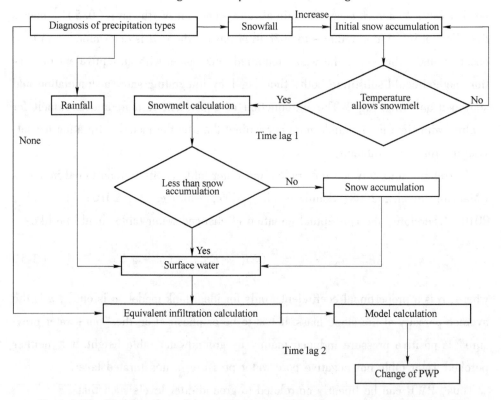

Figure 3.12 Workflow chart of the modified tank model with respect to the original model are highlighted in blue including time lags from snow accumulation/snowmelt and infiltration. It includes diagnosis of precipitation types, calculation of snow accumulation and melt, equivalent infiltration calculation, and calculation of change of PWP in a slip surface

3.3.2 Simpler approximations of slope hydrology

It assumes that quaternary deposits control the hydraulic properties of the tank model (tank interior with soil/rock in Figure 3.10). The fractured limestone and dolomite control the water flow from higher to lower elevations (groundwater inflow and drainage in Figure 3.10). The marly Kössen Beds are treated as impermeable layers (thin, low porosity and high normal stress above). As this is a regional groundwater table estimation, the modified tank model can be used to simulate the groundwater table changes induced by precipitation. Surface runoff flow resulting from snowmelt and heavy rainfall is ignored as (1) the slope angle is less than 15°, (2) the cumulative snowpack is no more than 70cm during monitoring days and (3) the infiltration rate of the slope in quaternary deposits and on carbonates is relatively high. Freezing effects on infiltration is ignored as (1) ground sealing by freezing is presumably not an issue since the bottom temperature of snow (BTS) is close to 0℃ underlain by a warmer subsoil in addition to high permeable subsoil. (2) Snow accumulation during winters and winter rainfall precipitation prevent effective cooling of the ground. Due to the all-year humid climate, the rapid drainage of water in the permeable underground and the deep-seated nature of the slope movement, evapotranspiration is not explicitly considered.

3.3.3 Determining the parameter of PWP calculation in the modified tank model

3.3.3.1 Correlation analysis between surface water and pore water pressure

The study split the data set into 18 months (excluding snowmelt and months of data losses) to evaluate the correlation coefficients. The IRS/AP (infiltration of rainfall (snowmelt)/absolute pore water pressure) correlation coefficient is usually smaller than the IRS/DP (infiltration of rainfall (snowmelt)/daily change of pore water pressure) correlation coefficient by 0.2~0.3. There is almost no correlation (IRS/RP) (infiltration of rainfall (snowmelt)/relative pore water pressure) and it was roughly lower than that (IRS/AP) at about 0.2~0.25. Table 3.3 and Table 3.4 show the results in heavy rainfall period (July, 2009 & August, 2009). 1~2 days response time (IRS/DP) based on all the monitoring data has a higher correlation coefficient of about 0.6~0.75 than shorter or longer response times. For monthly RS over 85 mm the response time is more likely 1 day (highest correlation coefficient); the correlation coefficient arrived peak with 2 days response time if monthly RS is between 20mm and 85mm; RS less than 20mm would delay the response time to 3 days for highest correlation coefficient.

Table 3.3 Correlation with IRS for the Aggenalm Landslide, July 2009

Response time/d	0	1	2	3	4
Correlation(AP)	-0.3863	0.0841	0.4157	0.4519	0.3256
Correlation(RP)	0.2072	0.0291	0.0268	0.1933	0.0615
Correlation(DP)	0.0234	0.7296	0.5095	0.0092	-0.1373

Table 3.4 Correlation with IRS respective for August 2009

Response time/d	0	1	2	3	4
Correlation(AP)	-0.1787	0.3334	0.4962	0.4190	0.3705
Correlation(RP)	0.0065	0.0666	0.1386	0.2047	0.1391
Correlation(DP)	0.0531	0.7427	0.3644	0.1937	-0.1787

On the basis of three years of monitoring of the rainfall, temperatures and pore pressures at a reactivated landslide in Bavaria, the relationships between IRS and DP are analyzed. Most previous attempts are based on the relationship of IRS and AP. The investigation of DP by IRS indicated they had a better linear correlation compared to AP by IRS and RP by IRS. Every landslide has a lowest water table (LWT). Since the AP=LWT+DP, the sensitivity of denominator by molecule IRS/(LWT+DP) is worse than (IRS/DP); while the RP (accumulation of DP), as time goes by, the sensitivity of denominator by molecule IRS/RP will reduce. For example the third day $IRS_3/RP_3(DP_1+DP_2+DP_3)$ is not so good as IRS_3/DP_3 (IRS_i means the intensity of ith day's RS; RP_i and DP_i are the ith day's values). When considering drainage process, DP is definitely an effective index. For instance, subsequent to a big RS event, the pore pressure is rising. Without new RS events in the next days, the AP may still rise but at a lower velocity or may even drop because of drainage. The same is true if a recurrent RS event appears while AP declines within the response time of RS. Therefore, the index AP is perhaps not so clear even producing variance because of the bigger denominator and impact from early days' DP_i. It should be noted that this study has examined only one in situ project which could be affected by external factors. Maybe for the landslide itself in drainage stage, calculation of DP is not so precise. But using DP can indeed reduce the impact from the problems mentioned above. On the other hand, response time seems to depend on the monthly RS. Increasing monthly amounts of RS can cause shorter response time possibly due to enhanced infiltration. Firstly, the heavier IRS will increase the water content faster in the same time especially in the beginning stage. So the response time is shorter facing bigger RS. Secondly, higher monthly amount of RS keeps the water table always in a higher level. And pores of soil are less and easily show the according positive pore pressure. Adversely, if the water table was very low, soil would absorb water for filling in its pores firstly not showing the positive pressure.

3.3.3.2 Correlation between infiltration, drainage, and daily change of pore water pressure

In order to determine an appropriate value of α' for the monitoring location on the Aggenalm Landslide, the study uses 13 months of data as training data to fit the relationship between equivalent rainfall and ΔPWP (Figure 3.13 and Figure 3.14).

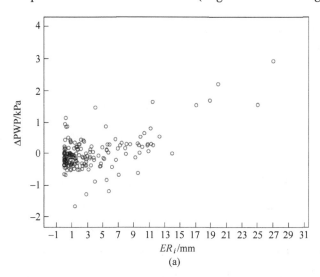
(a)

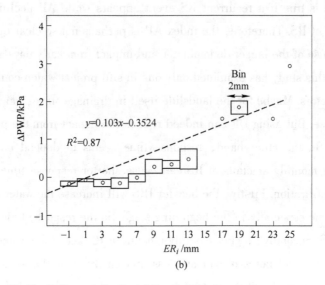

Figure 3.13 Daily equivalent rainfall and daily change of pore water pressure
(a) Daily equivalent rainfall ER_i versus daily change of pore water pressure ΔPWP_i in absolute values for 13 months (Sep. 2009~Feb. 2010 and May. 2010~Nov. 2010); (b) ΔPWP_i has been aggregated in bins of mean values for discrete steps of daily equivalent rainfall (mean+1 sigma error)

The linear relationship between daily change of pore water pressure (ΔPWP_i) and daily equivalent rainfall (ER_i) for absolute data is shown in Figure 3.13(a). However, this does not produce a functional link between ΔPWP_i and ER_i. We consider using the mean value of daily change of pore water pressure given for certain daily equivalent rainfall such as in Bin (Figure 3.13(b)) to replace the data of the same width (Figure 3.13(a)) (Freedman et al., 1998). The result shows the change of PWP_i as

$$\Delta PWP_i = \alpha' ER_i - \beta \qquad (3-8)$$

where, α' is 0.103kPa/mm, thus relates rainfall to pore pressure increase and β (−0.3524kPa) is the average decrease of pore water pressure by drainage. Thus at a day without infiltration by snowmelt and rainfall the pore water pressure drops by 0.35kPa, i.e., the water column drops by 35mm. According to the tank model theory, β, as a constant, is quite rough. The original tank theory demonstrates that the decrease of pore water pressure rate depends on the current pore water pressure (Michiue, 1985; Ohtsu et al., 2003; Takahashi, 2004; Takahashi et al., 2008; Xiong et al. , 2009; Uchimura et al. , 2010). In reality, the relationship can only be calculated by monitoring an extended period without infiltration. As shown in Figure 3.14(a), the observation of PWP is within 48 days without rainfall input which means these processes only include the information of drainage combined with groundwater supply for this test point. The relation between PWP_{i+1} and PWP_i without rainfall infiltration is shown in Figure 3.14(b) and Equation (3-9).

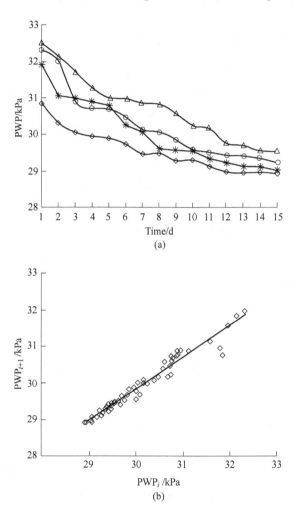

Figure 3.14 Drainage, and daily change of pore water pressure
(a) Observation of PWP vs. time for four fifteen-day-long periods without rainfall or snowmelt. Number of samples: $n = 48$; (b) PWP_i vs. PWP_{i+1} ith day of PWP correlates to $i+1$th day of PWP for four fifteen-day-long periods without rainfall or snowmelt. Number of samples: $n = 48$

$$\text{PWP}_{i+1} = a\text{PWP}_i + b \tag{3-9}$$

where, a and b are fitted coefficients.

Thus, calculation of ΔPWP_i could be rewritten as:

$$\Delta\text{PWP}_i = \alpha'(ER_i + ES_i) + ((a-1)\text{PWP}_i - b) \tag{3-10}$$

3.3.4 Snowmelt calculations in the modified tank model

3.3.4.1 Diagnosis of precipitation types

A threshold temperature under which the precipitation falls as solid snow is a key factor for a snow accumulation model. However, diagnosis of precipitation is difficult, and there are no parameters with which the type of precipitation can be determined for certainty (Wagner, 1957; Bocchieri, 1980; Czys et al., 1996; Ahrens, 2007). Until now, the most common approach is still to derive statistical relationships between some predictors and different precipitation types (Bourgouin, 2000). Therefore a statistical model is selected based on hundreds of observation samples in Wajima Japan, between 1975 and 1978 to estimate precipitation types (Matsuo and Sasyo, 1981). The threshold of relative humidity calculated by T_d (daily average temperature) is as follows:

$$RH_t = 124.9 e^{-0.0698 T_d} \tag{3-11}$$

If the real relative humidity RH is smaller than RH_t, the precipitation is usually snowfall (Häggmark and Ivarsson, 1997).

3.3.4.2 Snowmelt model

One of the most popular methods employed to forecast snowmelt is to correlate air temperature with snowmelt data. Such a relation was first used for an Alpine glacier by Finsterwalder (1887) and has been extensively applied and further refined since then (Kustas et al., 1994; Rango and Martinec, 1995; Hock, 1999, 2003). Recently, the most widely accepted temperature-index model is that of Hock (2003). The approach of daily melt assumes the form:

$$M' = f_m(T_d - T_0) \tag{3-12}$$

where, T_0 is a threshold temperature beyond which melt is assumed to occur (typically 0℃), and f_m is a degree-day factor. Widely used empirical f_m is suggested here (e. g. , Braun et al. , 1994; Hock, 2003), which is decided according to canopy cover of one area in percent, beginning time of snowmelt, etc. . In this study, the degree-day factor is calculated by

$$f_m = 2.92 - 0.0164F \qquad (3\text{-}13)$$

where, F is canopy covers of objective area in percent (Esko, 1980).

3.4 Results

3.4.1 Performance of modified tank model in heavy rainfall season

As shown in Figure 3.15, the modified tank model and original tank model considering no time lag are used to estimate the change of PWP in summer. Both the original and modified tank model do a reasonably good job at estimating changes in PWP during summer. The original model, however, generally overestimates the PWP curve. The modified model matches the measurement curve better due to the infiltration time lag 2.

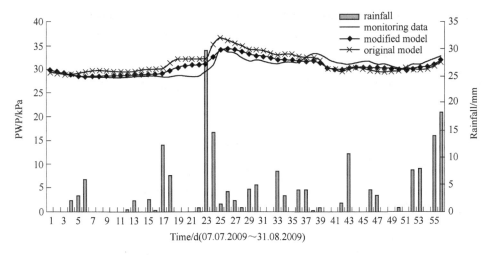

Figure 3.15 Estimation of change of PWP using the modified tank model (snowmelt + time lag 1+2) and the original tank model during summer (snow free) (07.07.2009~31.08.2009)

3.4.2 Performance of modified tank model in snowmelt season

The original model without snow accumulation and snowmelt does a poor job at estimating PWP during spring, as the change of PWP missing the accumulation time lag 1 caused by the original model to overestimate PWP from the day 12~33. The modified tank model much better reflects the peak of snowmelt ($33^{th} \sim 37^{th}$ day) and matches the measurement curve well in consideration of time lag 1. The deviation derives from the naturally limited accuracy of snow accumulation and snowmelt models (Figure 3.16).

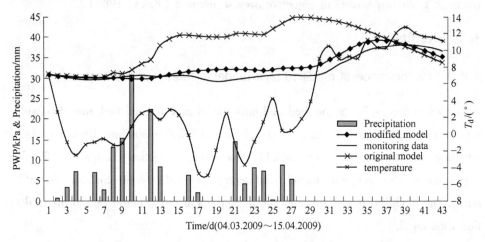

Figure 3.16   Estimation of change of PWP using our modified tank model (snowmelt + time lag 1+2) and original tank model in snowmelt season (04.03.2009~15.04.2009)

3.4.3 Performance of modified tank model throughout the monitoring period and error analysis

Error analysis in Figure 3.17 quantifies the model's performance. The Figure 3.17 indicates evaluation index of original and modified tank model including correlation, root

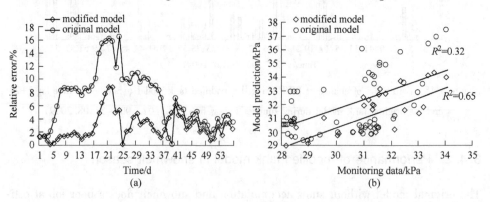

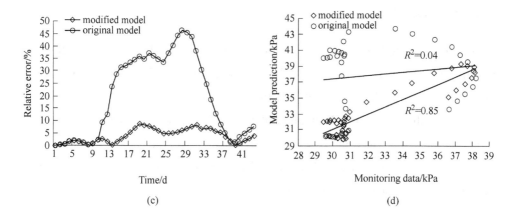

Figure 3.17 Evaluation of original and modified tank model
(a) Correlation between measurements and original/modified tank model during a 54-day rainfall period ($n=54$) Root mean square errors (RMSE) for the original and modified models are 1.9 and 0.97 respectively (07.07.2009~31.08.2009); (b) Correlation between measurements and original/modified tank model in snowmelt period ($n=47$). Root mean square error (RMSE): Original model 5.4/Modified model 1.3 (07.07.2009~31.08.2009); (c) Relative error of original and modified tank model in summer ($n=54$) (04.03.2009~15.04.2009); (d) Relative error of original and modified tank model during spring ($n=47$) (04.03.2009~15.04.2009)

mean square error (RMSE), and relative error. As shown in Figure 3.18, the modified tank model simulated the PWP levels in whole monitoring period.

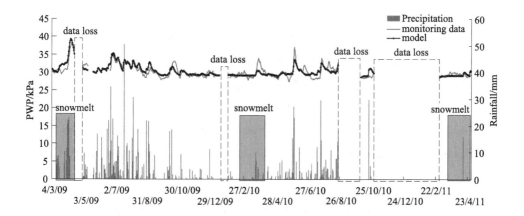

Figure 3.18 Simulation of PWP using the modified tank model throughout the monitoring period including snowmelt and heavy rainfall events (04.03.2009~23.04.2011)

3.5 Discussions

In order to evaluate the performance of the modified tank model with respect to heavy rainfall and snowmelt, it introduces the standard Nash-Sutcliffe efficiency (NSE) which is the most widely used criterion for calibration and evaluation of hydrological models with observed data (Nash and Sutcliffe, 1970). NSE is dimensionless and is scaled onto the interval [inf. to 1.0]. NSE is taken to be the 'mean of the observations' (Murphy, 1988) and if NSE is smaller than 0, the model is no better than using the observed mean as a predictor.

3.5.1 Performance of modified tank model in heavy rainfall season

The modified tank model describes the fluctuation of PWP reasonably well, especially during heavy rainfall days such as 23th to 26th day (43mm) and 51th to 55th day (45mm) (Figure. 3.15). The relative errors in Figure. 3.17(a) are less than 3% and 4% during these days. Dry periods (such as 2nd to 7th day and 17th to 21st day) agree with PWP measurement, with a relative error of 2%~9% as shown in Figure 3.17(a). The low water content of the landslide materials during the dry season appears to reduce the infiltration rates (Fredlund and Xing, 1994; Schaap and Van Genuchten, 2006). And PWP levels increase very slowly or not at all during these periods. As a result, the relative error of our modified model is slightly higher than that during wetter intervals. Compared with the original model, the modified model better represents PWP monitoring data. Figure 3.17(b) indicates a higher linear correlation between measurements and modified tank model with 0.65 (RMSE = -0.97) than the original tank model with 0.29 (RMSE = -1.9). The NSEs of the original tank model and the modified tank model during the heave rainfall season are -0.09 and 0.63 respectively. This means the standard original tank model is no better than the "mean of the observations" while the modified tank model has a significantly higher explanatory power.

3.5.2 Performance of modified tank model in snowmelt season

It found that a better correlation between measurements and the modified tank model with 0.86 (RMSE -0.97) than the original tank model in which all precipitation was assumed to be rainfall and snowmelt was not considered with 0.04 (RMSE = −5.4) during snowmelt period. It has to be pointed out that the snowmelt estimation is still not very precise, as the temperature-index model is relatively simple (Garen and Marks, 2005; Herrero et al., 2009; Lakhankar et al., 2013). Also, surface runoff is not considered due to the high permeability of surface deposits. The modified tank model, however, provides a useful estimation of increased PWP in creeping landslide masses several tens of meters deep. The NSEs of the original tank model and modified tank model during the snowmelt season are −5.95 and 0.75 respectively, which emphasizes the performance of the modified tank model.

3.5.3 Highlights of the modified model

Compared to the simple tank model, the modified tank model improves the prediction ability by introducing the equivalent infiltration method to reduce the infiltration time lags. Compared to the traditional deterministic methods (Fredlund and Xing, 1994; Chen and Young, 2006; Schaap and Van Genuchten, 2006; Weill et al., 2009), the modified tank model just needs historical monitoring data and does not need to consider uncertainties of material properties. Compared to the recent multi-tank model research (Ohtsu et al., 2003; Takahashi, 2004; Takahashi et al., 2008; Xiong et al., 2009), the modified tank model does not require complicated algorithms and several observation boreholes to optimize the parameters. It is a straightforward approach. The model integrates the snow accumulation/-melt model which is hardly considered in other tank model researches. A flexible approach is presented since the model can simulate the groundwater table at least two years continuously without obvious accumulative error unlike permeability-based numerical models or optimization parameter-based models need refreshment at times (Takahashi et al., 2008; Xiong et al., 2009).

3.5.4 Drawbacks and limitations

The naturally inevitable drawback for any "empirical model" is that it is physically not explicit. The presented model would need further adjustments for permafrost regions, with heavily frozen soils, for very steep slopes, with significant surface runoff and for very heterogeneous slopes, with complex fractured rock masses. However, it seems well suited for large mountain landslides on moderately inclined slopes in alpine conditions with significant snow accumulations.

3.6 Conclusions

Pore water pressure is one of the important dynamic factors in deep-seated slope destabilization and our modified tank model could help to anticipate critical states of deep-seated landslide stability a few days in advance by predicting changes in pore water pressure. In this chapter, a modified tank model is proposed for estimation of increased pore water pressure induced by rainfall or snowmelt events in a deep-seated landslide. Compared to the original tank model, the fluctuation of PWP is simulated more accurately by reducing the time lag effects induced by snowmelt and infiltration into a long path of deep-seated landslide. In this modified model, a statistical method based on temperature and humidity is used for diagnosis of precipitation types and a snowmelt model based on temperature index is integrated into it, also included an equivalent infiltration method which can describe the infiltration relative reliably is in the modified model to reduce their time lag effect.

4 Physical tank experiments on groundwater level controls of slopes with homogenous materials

Tank models are widely used for estimating groundwater levels in slopes. In this section, physical tank experiments are reported, indicating an evaluation of three typical conceptual tank models (simple tank model, surface runoff tank model, and lateral water flow supply tank model). Experimental results analyze groundwater tables were affected by infiltration time lags, surface runoff, and lateral water flow.

4.1 Introduction

In this study, three kinds of conceptual tank models for three typical slopes are considered. The distinct properties of the three tank models areas shown in Table 4.1.

Table 4.1 Characters of tank models

Type	Characters of applications
Simple tank model (Ishihara and Kobatake, 1979)	Simple model Assumed low slope angle; no surface runoff Negligible lateral water flow supply
Surface runoff tank model (Ohtsu et al., 2003)	Considering surface runoff Requires maximum infiltration rate test Less lateral water flow supply
Lateral water flow supply tank model (Takahashi et al., 2008; Xiong et al., 2009)	Increasing model complex Relatively high error Any slope angle

Simple tank model can be constructed and applied quickly, but the two typical limitations are a assumed low slope angle (one point for pore water pressure (PWP) represents the entire water table level) and a assumed high-porosity surface materials without considering surface runoff. Surface runoff tank models can overcome the limitations from surface material infiltration rates. However, maximum infiltration tests are necessary (Bodhinayake et al., 2004). The lateral water supply tank model can apply for any slope angle, but the complicated structure produces a higher systematically cumulative error. In this study, physical tank experiments are employed to investigate how infiltration time lag, maximum infiltration rate, and lateral water flows affect the estimation of pore water pressure (Figure 4.1).

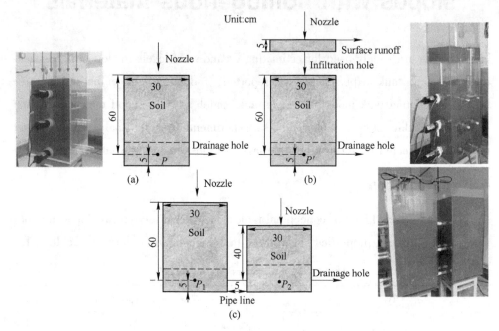

Figure 4.1 Physical tank model experiments
(a) Simple tank model. P is pore water pressure sensor; (b) Surface runoff tank model. P' is the PWP sensor;
(c) Lateral water flow supply tank model. P_1 and P_2 are PWP sensors

4.2 Methods

4.2.1 Test setup and testing materials

4.2.1.1 Physical tank models

A series of physical tank model systems made of plexiglass with soil samples to simulate the slope mass of a homogenous hillslope (Figure 4.1). A plate squeezes the soil layer with a force of 120N to form the slope mass (squeeze soil one time per adding 5cm soil layer) (Figure 4.2).

Figure 4.2 Soil filling into the tank model. Every 5cm depth soil layer is pressed by the flat

4.2.1.2 Rainfall simulator

A water-pump supplies water resources through one or two nozzles (uniformity coefficient is around 0.87) for rainfall simulation. A flow meter between pump and nozzles controls the rainfall intensity (10~250mL/min). The rainfall simulator is shown in Figure 4.3. A water storage tank is used for the water supply. A pump can improve the water pressure before rainfall production. The purpose of the flowmeter is to adjust flow rate for producing different rainfall intensities. The spray nozzles can produce the uniform rainfall. For the rainfall simulations, the rainfall intensity test and uniform degree test of rainfall are necessary before the experiments.

Rainfall intensity test: the simple tank (300mm×300mm×300mm) is used for rainfall intensity testing. The collected rainfall in unit time is compared to the calculating rainfall depending on the flowmeter. The test range of the flow rate of the flowmeter is 15~120mL/min and the increment of the test value is 15mL/min. Each test lasts 0.5 hr. The test result is shown in Figure 4.4.

Uniform degree test of rainfall: generally the uniformity coefficient of a stable rain-

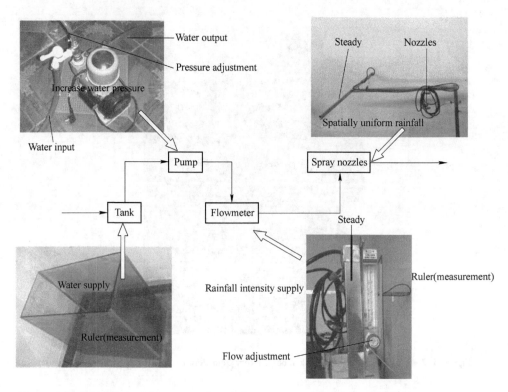

Figure 4.3 Rainfall simulator systems. It includes a water supply (tank with ruler), a water output (pump), a rainfall intensity adjustment (flowmeter), and a rainfall output (spray nozzles)

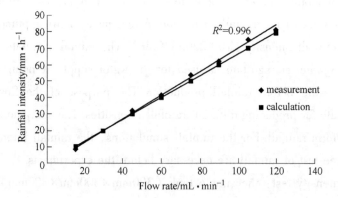

Figure 4.4 Rainfall intensity simulation test (rainfall intensity measurement vs. rainfall intensity calculation under different flow rates input)

fall simulation should be bigger than 0.8 (Lora et al., 2016). The uniform degree of rainfall can be calculated by Equation (4-1) as follows:

$$k = 1 - \sum_{i=1}^{n} \frac{|x_i - x|}{nx} \qquad (4\text{-}1)$$

where, k is the uniformity coefficient; x_i is the rainfall of measurement positions; x is the average rainfall of measurement positions; n is the number of measurement positions.

In this test, four measuring glasses are randomly placed in the rainfall zone. The simulated rainfall intensities are 25mm/h, 45mm/h, and 65mm/h in three rainfall events and every rainfall event lasts 30min. The uniformity coefficients are 0.84, 0.88, and 0.90, respectively.

4.2.1.3 Pore water pressure and drainage records

Pore water pressure transducers (Model number CYY2, Xi'an Weizheng Technology Corp., Ltd, Xi'an, China) have a diameter of 3cm, a height of 1.6cm, measuring range +10kPa, deviation 0.2%. They are used to record the PWP (Figure 4.5).

Figure 4.5 Pore water pressure transducers

Pore water pressure transducers calibration is necessary before the experiments. A tank (300mm×300mm×700mm) is employed for the calibration of the pore water pressure transducers. The transducers are placed in the bottom of the tank. Every time 3cm hight water table is added into the tank and the monitoring value of transducers is recorded (the output of transducers is electric current). The calibration results are shown in Figure 4.6.

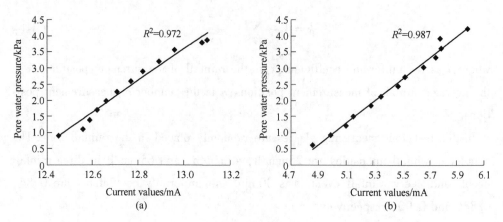

Figure 4.6 Pore water pressure transducers (P_1 and P_2) calibration results

(a) P_1(pore water pressure values vs. current values); (b) P_2(pore water pressure values vs. current values)

A channel on the right bottom of the physical tank model is used to calculate the drainage. A measuring glass collects the drainage and a camera records the change of the water in glass (Figure 4.7).

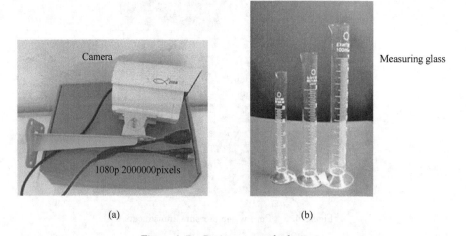

Figure 4.7 Drainage records devices

(a) employed camera (for recording the changes of water hight in measuring glass);

(b) the measuring glass (for collecting the drainage)

4.2.1.4 Data collection and software

In this test, data acquisition system (Figure 4.8) (CK01L0R-C20 type) with RS485 communication interface is used for data collection, which can do a long-distance data transmission. The system uses a MODBUS-RTU protocol and has a high data transmission stability and a versatility for example the multi-channel Analog input and 14 bit ADC precision. The collected object is electric current (0~20mA) which comes from the sensors.

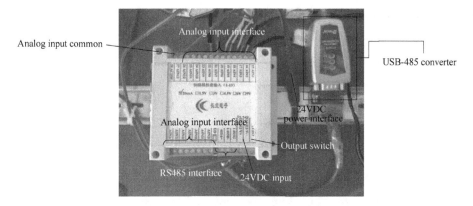

Figure 4.8 Data acquisition system. It includes a Analog input interface, a 24 VDC power interface, and a USB-485 converter

The software designed by VC++ displays real-time data and dynamic curves, whose interface is divided into three main parts (Figure 4.9). The left one shows the real-time water pressure value; the right part produces the monitor data graph; in the lower part, the original signals are displayed.

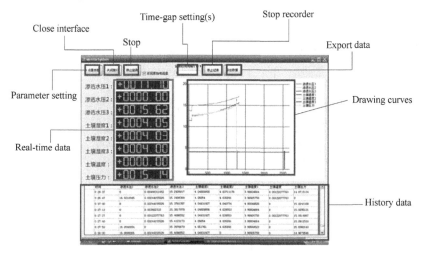

Figure 4.9 Software for displaying monitoring data. The functions include defining the units, producing the monitor data graph, recording data, and adjusting the monitoring time interval

4.2.1.5 Testing material

The soil used in experiments is relatively homogeneous from Ming Mountain, near the Yangtze Riverbank, Chongqing, China (Figure 4.10). Ming Mountain is located at the Three Gorges Reservoir Area, where the average annual rainfall is 1074.6mm and 70% of the annual rainfall occurs between May and September. Soil consists of quaternary alluvial materials.

Figure 4.10 Soil samples location. They are collected from the toe of Ming Mountain, near the Yangtze River Bank, Chongqing, China

The soil materials taken totally achieve 50kg. The materials are screened by a sieve soil. The soil's particle-size distribution curve is as shown in Figure 4.11. The density

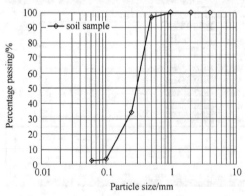

Figure 4.11 Particle-size distribution curves. 90% of particle-size concentrates in a range of 0.1~0.4mm

of materials is 1.82~1.85g/cm^3, while saturated density is 2.04~2.07g/cm^3.

4.2.2 Experiment procedures

Every test is conducted under similar initial conditions, such as geometry, material, moisture content, and initial groundwater level (PWP) (deviation±3%).

4.2.2.1 Simple tank experiment

The experiments include fixed and variable rainfall intensity-duration inputs for the hydrology calculations (Table 4.2). For example, the test 1 simulates 25mm/h (36min) rainfall event, while the test 4 simulates the rainfall event of 25mm/h (12min),65mm/h (12min),25mm/h (12min). The PWP sensor at the bottom of tank records the changes of PWP during the rainfall events. The measuring glass collects the drainage.

Table 4.2 Experiment arrangements of simple tank model

No.	Rainfall input-intensity (duration)	Output objects
1	25mm/h (36min)	P; drainage
2	45mm/h (36min)	P; drainage
3	65mm/h (36min)	P; drainage
4	25mm/h (12min),65mm/h (12min),25mm/h (12min)	P; drainage
5	25mm/h (24min)	P; drainage
6	45mm/h (24min)	P; drainage
7	65mm/h (24min)	P; drainage

4.2.2.2 Surface runoff tank experiment

These experiments are to investigate how the maximum infiltration rate limits the PWP by reducing rainfall infiltration (Table 4.3). For example, the test 1 simulates 25mm/h (24min) rainfall event, while the test 3 simulates the rainfall event of 65 mm/h (24min). The PWP sensor at the bottom of tank records the changes of PWP during the rainfall events. The measuring glass collects the drainage.

Table 4.3 Experiments of surface runoff tank model

No.	Rainfall input-intensity (duration)	Output objects
1	25mm/h (24min)	P'; drainage; surface runoff
2	45mm/h (24min)	P'; drainage; surface runoff
3	65mm/h (24min)	P'; drainage; surface runoff

4.2.2.3 Lateral water flow supply tank experiment

These experiments are to investigate how the lateral water flow affect the PWP in both tanks (Table 4.4). For example, the test 1 simulates 45mm/h (24min) rainfall event, while the test 2 simulates the rainfall event of 65mm/h (36min). The PWP sensors at the bottoms of both tanks record the changes of PWP during the rainfall events. The measuring glass collects the drainage.

Table 4.4 Experiments of lateral water flow supply tank model

No.	Rainfall input-intensity (duration)	Output objects
1	45mm/h (24min)	P_1; P_2; drainage
2	65mm/h (36min)	P_1; P_2; drainage

4.3 Results and analysis

4.3.1 Simple tank experiment

Figure 4.12 shows the PWP and drainage during the test 1~7 (Table 4.2). The PWP in the whole processes is divided into three stages: (1) initial stage without obvious increase; (2) increase stage considering drainage; (3) decrease stage without rainfall infiltration. It is found that the amount of rainfall affects the value and time of PWP peak. Simply, a high rainfall value means a short time lag and a high value of the PWP peak.

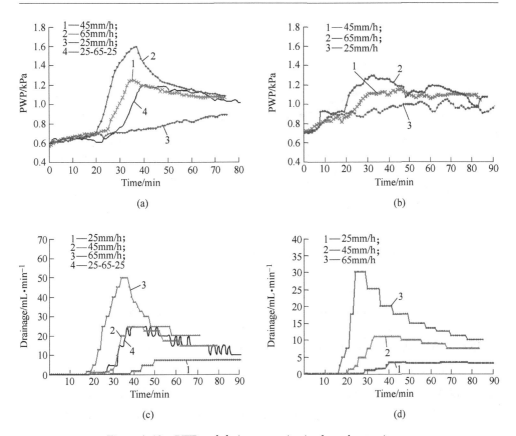

Figure 4.12 PWP and drainage rate in simple tank experiments

(a) PWP vs. time (test 1~4). Peak values of PWP vary from 0.7 to 1.6kPa; peak times of PWP vary from 36 to 40min; (b) PWP vs. time (test 5~7). Peak values of PWP are from 0.9 to 1.3kPa; peak times of PWP are from 30 to 40min; (c) Drainage vs. time (test 1~4). Peak values of drainage rates are from 10 to 50mL/min; peak times of drainage rates are from 30 to 50min; (d) Drainage vs. time (test 5~7). Peak values of drainage rates are from 5 to 30mL/min; peak times of drainage rates are from 24 to 40min

Figure 4.13(a) investigates the relation between cumulative rainfall and PWP. The initial stage is linear or constant and second increase stage could be a log function. The decrease stage involves the power or exponent function. Figure 4.13(b) demonstrates higher PWP causes a faster drainage than lower PWP. And the relationship between PWP and drainage rate is linear.

4.3.2 Surface runoff tank experiment

The distinct point between the simple and surface runoff tank experiment is the maximum infiltration rate limitation. The upper tank has a limited infiltration ability realized by a

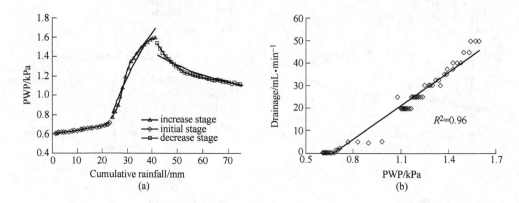

Figure 4.13 Relationship analysis of PWP, cumulative rainfall, and drainage
(a) PWP vs. cumulative rainfall (test 3). Degrees of correlations between PWP and cumulative rainfall in the initial, increase, and decrease stages are 0.93, 0.96, and 0.92, respectively;
(b) PWP vs. drainage (test 3). Linear correlation degree is 0.96

infiltration hole downward. In other words, if there is a heavy rainfall, the upper tank could drain some of the water as surface runoff. As shown in Figure 4.14(a), the PWP under the 45mm/h and 65mm/h rainfall events are lower than PWP under the same rainfall events during simple tank experiments (Figure 4.12(a)). The surface runoff in Figure 4.14(b) shows the infiltration threshold controlled the rainfall surface runoff. For the bottom drainage (Figure 4.14(c)), the results are similar to the simple tank model, except that the maximum infiltration reduces the amount of drainage. Figure 4.14(d) shows

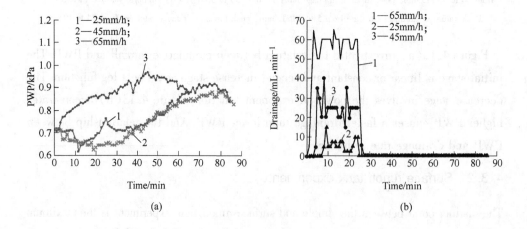

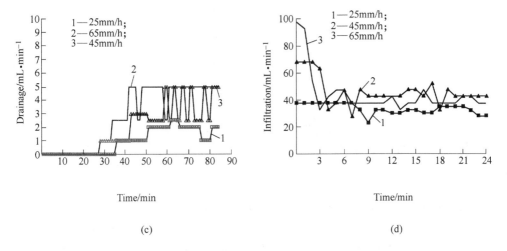

Figure 4.14 Monitoring data from P', surface runoff and drainage hole (surface runoff tank model): (a) PWP (P') vs. time (test 1-3). Peak values of PWP are from 0.85 to 0.95kPa; peak times of PWP are 45, 75, 80min for 65mm/h, 45mm/h, and 25mm/h rainfall events, respectively; (b) Surface runoff vs. time (test 1-3). During the rainfall periods the drainage rates are 60mL/min, 25mL/min, and 7mL/min for 65mm/h, 45mm/h, and 25mm/h rainfall events, respectively; (c) drainage vs. time (test 1-3). Peak values of drainage rates are 2mL/min, 5mL/min, and 5mL/min for the 25mm/h, 45mm/h, and 65mm/h rainfall events, respectively; peak times of drainage rates are 40min, 60min, and 63min for 65mm/h, 45mm/h, and 25mm/h rainfall events, respectively; (d) Infiltration vs. time (test 1-3). The maximum infiltration limits the infiltration rate to 38mL/min for the 25mm/h, 45mm/h, and 65mm/h rainfall events

that the maximum infiltration is about 38mL/min.

4.3.3 Lateral water flow supply tank experiment

Lateral water flow supply complicates the calculation of groundwater table especially coupling the infiltration time lags. In Figure 4.15(a), the P_2 of right (lower) tank model firstly begin to increase due to the water supply from left (higher) tank model. The left tank model as the water supplier mostly affects the right one, although in the beginning the right one could offer some water to the left one conversely (as shown in Figure 4.1(c), right tank's short vertical infiltration path could produce a higher groundwater table than the left tank in the beginning).

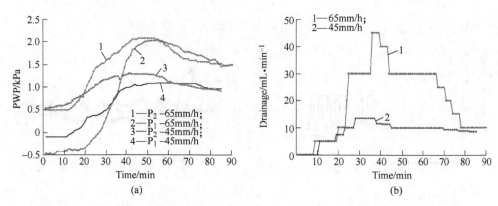

Figure 4.15 Monitoring data from PWP sensors and drainage hole (water flow supply tank model) (a) PWP (P_1 and P_2) vs. time (test 1-2). P_1 indicates increased PWP are (-0.1 to 1)kPa and (-0.5 to 2)kPa for the 45mm/h, and 65mm/h rainfall events; P_2 indicates increased PWP are (0.5 to 1.25)kPa and (0.5 to 2)kPa for the 45mm/h, and 65mm/hr rainfall events; peak times of PWP of all the P_1 and P_2 are between 40min and 50min; (b) Drainage vs. time (test 1-2). Peak values of drainage rates are 13mL/min and 45mL/min for the 45mm/h, and 65mm/h rainfall events; peak times of drainage rates are around 35min for both 45mm/h, and 65mm/h rainfall events

4.4 Conclusions

Changes of PWP are controlled by the balancing among the rainfall infiltration, water flow supply, and the drainage. The relationships between PWP on the bottom of tank, drainage, and rainfall based on three kinds of physical models (simple tank model, surface runoff tank model, and water flow supply tank model) are investigated. Drainage processes under different rainfall events are also deciphered. Some concolusions are as follows:

(1) The amount of rainfall affects the value and time of PWP peak. Simply, a high rainfall value means a short time lag and a high value of the PWP peak.

(2) Infiltration threshold of surface soil controls the rainfall surface runoff and maximum infiltration.

(3) Lateral water flow from a higher part to a lower part of a slope system can fast improve the PWP of the lower part.

5 Physical tank experiments for estimation of groundwater considering slope structure controlling affection

5.1 Introduction

The conceptual tank model is employed to predict groundwater table changes, mainly focusing on homogenous materials (Michiue, 1985; Ohtsu et al., 2003; Takahashi, 2004; Takahashi et al., 2008; Xiong et al., 2009). For heterogeneous materials, the conceptual tank model still needs to be validated for predicting the groundwater table changes. In this section, physical tank experiments were continuously carried out for the pore water pressure calculation of four complex slope structures. These include a coarse-fine material slope, a slope including a fine layer, a slope including an obvious fracture, and a slope in interaction with a river.

5.2 Different typical geological condition of landslides

5.2.1 A coarse-fine material slope

A coarse-fine material slope comprises coarser material in the upper part due to being weathered and finer material in the lower part considering the movement of sediment. Some cases belonging to this classification include the Aggenalm Landslide in Bayern, Germany (Figure 5.1), and Lichtenstein-Unterhausen Landslide in Swabian Alb, Germany (Thiebes et al., 2014). The infiltration process is generally described in Figure 5.2.

5.2.2 A slope including a fine layer

A slope including a fine layer could produce a perched water table in the main slope mass considering the very low permeability of the fine layer. The infiltration of precipitation usually has a huge time lag in the material. Therefore, there is a longer re-

5 Physical tank experiments for estimation of groundwater considering slope structure controlling affection

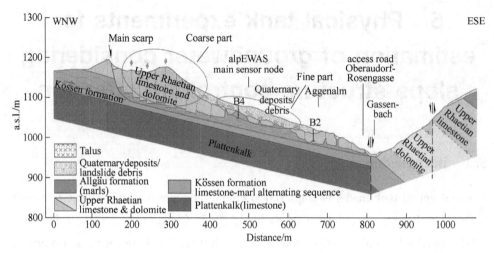

Figure 5.1　Geological profile of the Aggenalm Landslide in Bayern, Germany. The upper part consists of weathered limestone and dolomite (coarse material); the lower part is quaternary deposits which have a low porosity (fine material) (after Festl 2014)

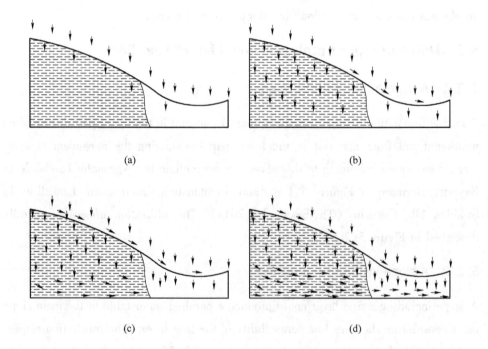

5.2 Different typical geological condition of landslides · 67 ·

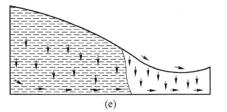
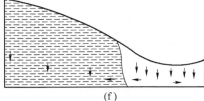

(e)   (f)

Figure 5.2 Precipitation on a coarse-fine material slope
(a) Start of rainfall; (b) Permeability in coarse part is higher than that in fine part. The water can infiltrate the coarse material easily, while the fine material can produce the surface runoff due to the low infiltration rate; (c) Lateral water flow moves fast in coarse material; (d) Lateral water flow moves from the coarse material to the fine material; (e) Reduction of lateral water flow and infiltration; (f) Lateral water flow in fine material moves towards both sides

Coarse;   Fine

sponse time about changes in the groundwater table. A typical case is the Ventnor Undercliff Landslide, Isle of Wight, England (Figure 5.3) (Moore et al., 2006). The infiltration process is generally described in Figure 5.4.

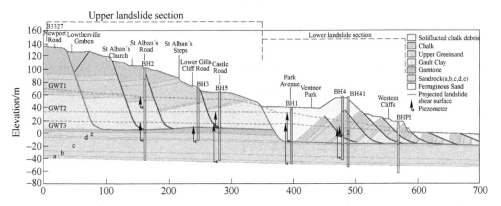

Figure 5.3 The Ventnor Landslide ground model. The Gault clay (light blue part) can be treated as a fine material which has a low permeability (Moore et al., 2006)

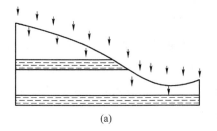
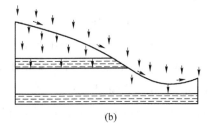

(a)   (b)

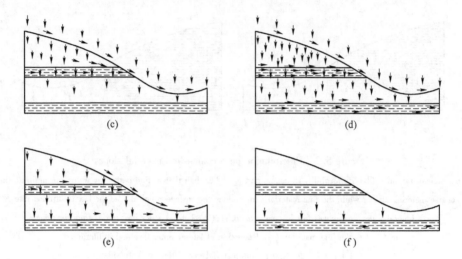

Figure 5.4 Precipitation on a slope including a fine layer

(a) Rainfall beginning; (b) Permeability in fine layer is lower than that in the coarse material and the infiltration becomes slow; (c) Perched water table produces around the fine layer part because of low permeability. And the lateral water flow moves (d) Water infiltration begins towards down; (e) Increase of groundwater table appears even the rainfall stops; (f) Water flow exits even when rainfall stops

≡≡≡ Fine; ☐ Coarse

5.2.3 A slope including an obvious fracture

A case of a slope including obvious fractures is located in Satsuma, Japan (Figure 5.5) (Wakizaka, 2013). The infiltration process is generally described in Figure 5.6.

5.2.4 A slope in interaction with a river

River-groundwater interaction in a river-slope system under a rainfall event is common in riverbank or basin areas (Doussan et al., 1998; Rosenberry and Healy, 2012). The estimation of groundwater is complicated, resulting from groundwater-river interactions which could relate to the permeability, hydraulic gradients, and hydro geological properties in slope-river system (Trémolières et al., 1993; Doussan et al., 1998; Rosenberry and Healy, 2012; Gollnitz, 2003; Mutiti and Levy, 2010). A case in Figure 5.7, located on the South Platte River, north of Denver, Colorado, USA (Rosenberry and Healy, 2012). The infiltration and water exchange process can be generally described in Figure 5.8.

5.2 Different typical geological condition of landslides

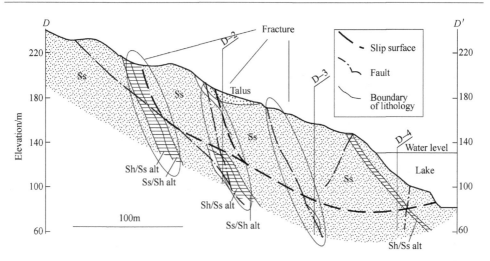

Figure 5.5 Geological profile of a landslide located in Satsuma, Japan. The fractures along faults can be considered with high permeability in the slope mass (after Wakizaka, 2013)

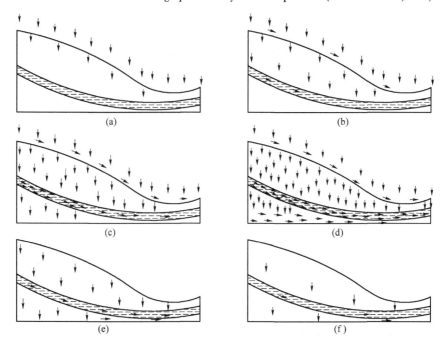

Figure 5.6 Precipitation on a slope including an obvious fracture

(a) Rainfall beginning; (b) Infiltration in fine layer and the surface runoff produces; (c) Fast water flow produces in fracture layer because of high permeability; (d) Water infiltration penetrates the fracture layer downwards and the groundwater table produced is not too high; (e) Groundwater table reduces when rainfall stops; (f) Water flow and infiltration reduce when rainfall stops

⸺ Coarse;   ☐ Fine

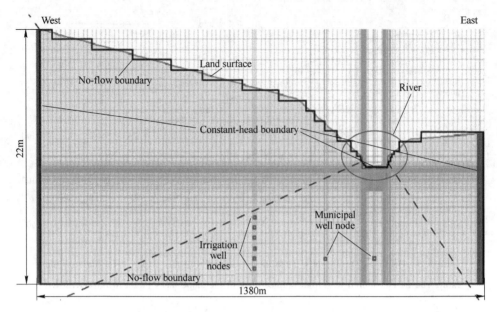

Figure 5.7 River-groundwater slope on the South Platte River, north of Denver, Colorado, USA (Rosenberry and Healy, 2012)

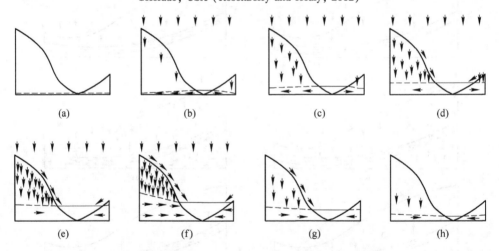

Figure 5.8 River-groundwater interaction under rainfall event

(a) Initial state; (b) Groundwater level raised by rainfall infiltration and river supply; (c) More rainfall and river supply and overland flow produced; (d) High groundwater table due to continuous rainfall and the river supply; (e) Groundwater conversely supplies the river level; (f) Increased groundwater accelerates the water flow supply to the river; (g) Overland flow and rainfall infiltration reduce; (h) Recovery to the initial state

5.3 Physical tank experiments

A series of physical multi-tank experiments are carried out by simulating groundwater table changes in consideration of four geological conditions. For information on rainfall simulator, monitoring system, soil sample collection, and devices calibration see chapter 4.

5.3.1 Tank experimental flume and materials

A series of physical tank model systems made of plexiglass with field soil simulate the slope mass of a representative hillslope. A plate squeezed the soil layer with a force of 120N to form the slope (squeeze soil one time per adding 5cm soil layer). A water-pump supplies water resources through two nozzles (uniformity coefficient is 0.91) for rainfall simulation. A flow meter between pump and nozzles controls the rainfall intensity (10~250mL/min). In the tank experiment, the materials chosen are as shown in Figure 5.9. The particle-size distribution curves are shown in Figure 5.9(c).

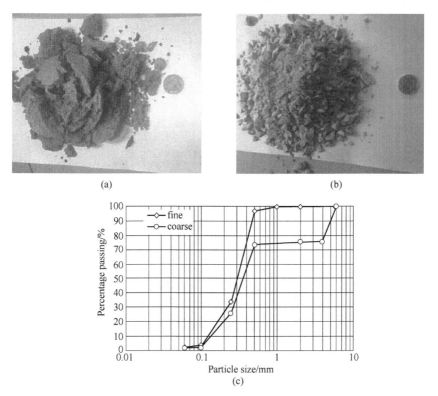

Figure 5.9 Materials used in physical tank experiments

(a) Fine material (mainly clay); (b) Coarse material (mainly sand, partly small gravel stones);

(c) Particle-size distribution curves (particle sizes (around 98%) of fine material are smaller than 0.4mm; particle sizes (4~8mm) of coarse material occupy 25%)

5.3.1.1 Coarse-fine material slope

The left tank, filled with coarse material, is used to simulate the higher coarse material part, while the right tank is filled with the fine material for simulation of the lower slope part. Rainfall is simulated by nozzles. Two PWP sensors are installed at the bottom of double tanks to monitor pore water pressure. The drainage of this system is realised through a drain hole(Figure 5.10).

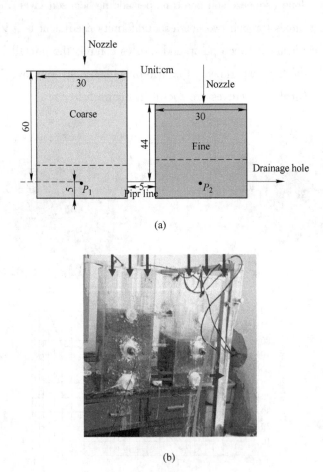

Figure 5.10 Coarse-fine material slope modelling device. The left tank filled with coarse material (hight of 60cm) is used to simulate the upper part of a slope; the right tank filled with fine material(hight of 44cm) is used to simulate the toe part of a slope
(a) Schematic diagram; (b) Real scene

5.3.1.2 A slope including a fine layer

The double tanks consist of coarse material as the main slope mass. In the horizontal direction, fine material is used to simulate low permeability material such as clay layers. Rainfall is simulated by nozzles. Two PWP sensors are installed at the bottom of double tanks for pore water pressure monitoring. The drainage of this system is realised through a drain hole (Figure 5.11).

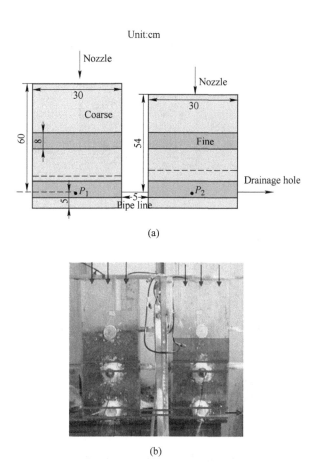

Figure 5.11 Modelling of a slope including fine layers. The left tank (hight of 60cm) is used to simulate the upper part of a slope; the right tank (hight of 54cm) is used to simulate the toe part of a slope; both of tanks are inserted with fine material layers (high of 8cm) (a) Schematic diagram; (b) Real scene

5.3.1.3　A slope including an obvious fracture

The double tanks consist of fine material as the main slope mass. Coarse material is employed to simulate high permeability material such as a fracture. Rainfall is simulated by nozzles. Two PWP sensors are installed at the bottom of double tanks for pore water pressure monitoring. The drainage of this system is realised through a drain hole (Figure 5.12).

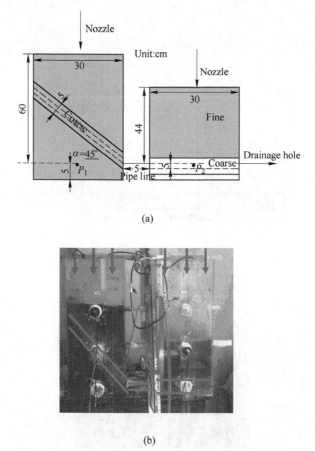

(a)

(b)

Figure 5.12　Obvious fracture-included slope modelling device. The left tank filled with fine material (hight of 60cm) is used to simulate the upper part of a slope; the right tank filled with fine material (hight of 44cm) is used to simulate the toe part of a slope; both of tanks are inserted with coarse material layer for a simulation of a fracture (high of 5cm)

(a) Schematic diagram; (b) Real scene

5.3.1.4 River-groundwater slope

The left tank consists of coarse material as the main slope mass. The left small tank filled with fine material works as a surface soil layer which has a maximum infiltration rate and a potential surface runoff function, while, the right tank is empty with a basic water table such as that of a river. Rainfall is simulated by nozzles. Two PWP sensors are installed at the bottom of double tanks for pore water pressure monitoring. The drainage of this system is realised through a drain hole (Figure 5.13).

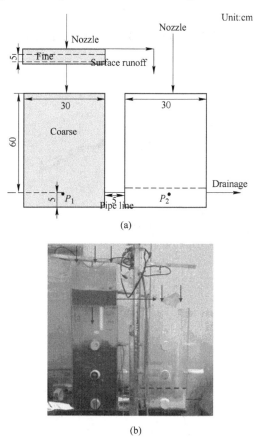

Figure 5.13 River-groundwater slope modelling device. The left tank filled with coarse material (hight of 60cm) is used to simulate the upper part of a slope; the left small tank filled with fine material (hight of 5cm) works as a surface soil layer which has a maximum infiltration rate and a potential surface runoff function; the right tank has a initial water table (hight of 6cm) is used for a simulation of a river
(a) Schematic diagram; (b) Real scene

5.3.2 Physical tank experimental outline

For each group of physical tank experiments, two rainfall events (45mm/h and 65mm/h intensity, 36min duration) are arranged and the observation time was 1 hr. Every test in each group was conducted under similar initial conditions, such as geometry, material, moisture content, and initial PWP (deviation ±3%).

5.4 Results and discussion

(1) Coarse-fine material slope.

Figure 5.14 shows the PWP of both tanks and the drainage rate of coarse-fine material slope experiment.

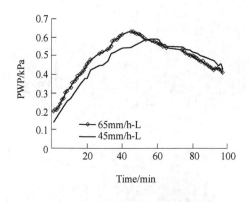

(a)

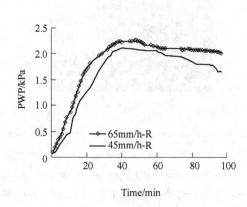

(b)

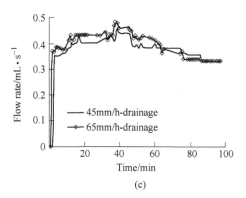

Figure 5.14 Pore water pressure of a coarse-fine material slope in physical experiments
(a) Left tank (P_1). The peak values of PWP are 0.6kPa/h and 0.5kPa for 65mm/h and 45mm/h rainfall events; the peak times of PWP are 45min and 55min for 65mm/h and 45mm/h rainfall events; (b) Right tank (P_2). The peak values of PWP are 2.2kPa and 2kPa for 65mm/h and 45mm/h rainfall events; the peak times of PWP are 40min for both 65mm/h and 45mm/h rainfall events; (c) Drainage. The peak values of flow rates are 0.45mL/s and 0.43mL/s for 65mm/h and 45mm/h rainfall events; the peak times of flow rates are 40min for both 65mm/h and 45mm/h rainfall events

Because the water infiltration and flow in coarse part are faster thanthat in the fine materials, the main water movement is towards the right tank. In other words, the inverse function from the slope toe (right) towards the upper part (left) is less. The main difference between left and right tank is just after the rainfall stops: in the right tank the changes of PWP has a longer time lag than that in the left tank.

(2) A slope including a fine layer.

Figure 5.15 shows the PWP of both tanks and the drainage rate of the fine layer included in the slope experiment.

In this situation, the water infiltration in the toe (right) part could be faster due to a shorter path than that in the left part. Plus the water flow supply from left tank, the PWP in right tank can achieve a high value. In the left part, the PWP values cannot increase too much because of the water flow to the toe part and a long time lag of infiltration, (The peak value is around 2kPa compared to around 6kPa in right tank). The obvious truth is the peaks of PWP in both tanks keep their high levels in a quite long time after the rainfall stop (36min). This reason is that two fine layers delay the water downwards.

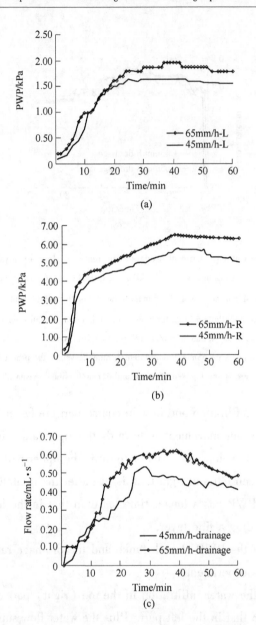

Figure 5.15 Pore water pressure of fine layer included slope in the physical experiments
(a) Left tank (P_1). The peak values of PWP are 1.9kPa and 1.6kPa for 65mm/h and 45mm/h rainfall events; the peak times of PWP are 38min and 25min for 65mm/h and 45mm/h rainfall events; (b) Right tank (P_2). The peak values of PWP are 6.4kPa and 5.4kPa for 65mm/h and 45mm/h rainfall events; the peak times of PWP are 40min for both 65mm/h and 45mm/h rainfall events; (c) Drainage. The peak values of flow rate are 0.6mL/s and 0.5mL/s for 65mm/h and 45mm/h rainfall events; the peak times of flow rate are 35min and 30min for 65mm/h and 45mm/h rainfall events

(3) A slope including an obvious fracture.

Figure 5.16 shows the PWP of both tanks and the drainage rate of obvious fracture-included slope experiment.

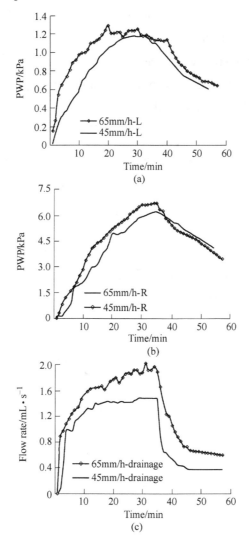

Figure 5.16　Pore water pressure of a slope including an obvious fracture in physical experiments (a) Left tank (P_1). The peak values of PWP are 1.2kPa and 1.1kPa for 65mm/h and 45mm/h rainfall events, respectively; the peak times of PWP are 20min and 30min for 65mm/h and 45mm/h rainfall events; (b) Right tank (P_2). The peak values of PWP are 6.8kPa and 6.1kPa for 65mm/h and 45mm/h rainfall events; the peak times of PWP are 35min for both 65mm/h and 45mm/h rainfall events; (c) Drainage. The peak values of flow rate are 1.8mL/s and 1.3mL/s for 65mm/h and 45mm/h rainfall events; the peak times of flow rate are 35min for both 65mm/h and 45mm/h rainfall events

In this situation, the PWP at the left tank bottom was much less because the most water flow along the main fracture (45 degrees fracture angle), (the peak value is no more than 1.2kPa). By contrast, the PWP (around 6kPa) in the right side can be supplied by the fast water flow through the main fracture. There is also no water flowing back from the right to left tank.

(4) A slope in interaction with a river.

Figure 5.17 shows the PWP of both tanks and the drainage rate of rive-groundwater slope experiment (coarse material).

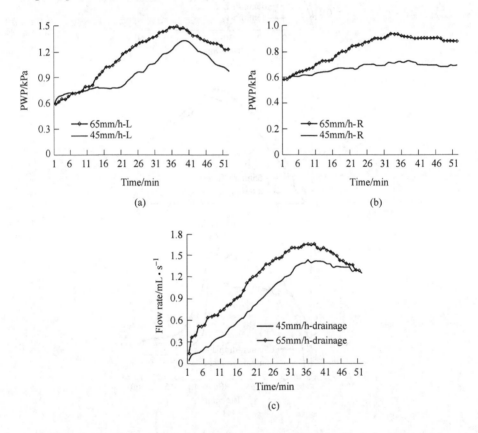

Figure 5.17 Pore water pressure of rive-groundwater slope in physical experiments
(a) Left tank (P_1). The peak values of PWP are 1.45kPa and 1.3kPa for 65mm/h and 45mm/h rainfall events; the peak times of PWP are 37min and 40min for 65mm/h and 45mm/h rainfall events; (b) Right tank (P_2). The peak values of PWP are 0.9kPa and 0.65kPa for 65mm/h and 45mm/h rainfall events; the peak times of PWP are 36min for both 65mm/h and 45mm/h rainfall events; (c) Drainage. The peak values of flow rate are 1.6mL/s and 1.4mL/s for 65mm/h and 45mm/h rainfall events; the peak times of flow rate are 38min for both 65mm/h and 45mm/h rainfall events

The right tank is the river part of the river-slope system. Compared to a porous material-soil, the increase of river level is not sensitive enough to rainfall. In other words, adding the same water can increase a higher groundwater table in soil than the water level in river.

5.5 Summary and conclusions

Physical tank experiments were continuously carried out for the pore water pressure of four complex landslide structures (a coarse-fine material slope, a slope including a fine layer, a slope including an obvious fracture, and a river-groundwater interaction slope). Some conclusions are as follows:

(1) For different types of slope structures, the groundwater changes are different under the same rainfall events. Using only one conceptual tank model for prediction of groundwater table of different types slopes may be not practical.

(2) For a coarse-fine material slope, the changes of PWP in fine material part has a longer time lag than that in the coarse material part. And the toe part (fine material part) usually has a much higher PWP than that in upper part (coarse material part).

(3) For a slope including a fine layer, the peaks of PWP in both parts (toe and upper) keep their high levels in a quite long time after the rainfall stop.

(4) For a slope including an obvious fracture, most of infiltration becomes the water flow along the main fracture. Thus, the upper part has a low PWP value due to the fracture carrying most water from this area, while the toe part has a high PWP value due to the supply from the upper part by the fracture.

(5) For a river-groundwater interaction slope, adding the same water can increase a higher groundwater table in soil than the water level in river.

6 Prediction of groundwater affecting deep-seated landslide quasistatic movement

Traditionally, in a viscous model of a landslide, changes of the normal stress caused by groundwater fluctuation and precipitation infiltration are regarded as a main factor to control the velocity of the movement. In this study, the variable weight of slope, and especially cohesion dependent on velocity, are necessary for the prediction of landslide movement. These considerations can reduce the cumulative prediction error of the landslide model. Examples of applications of the new viscous landslide model to the Ventnor landslide, in the Isle of Wight, southern England, and to the Utiku landslide in New Zealand were discussed. In both cases, a more successful calibration of the model was achieved, despite unavoidable uncertainties concerning the dates of occurrence of the slope movements. By introducing the standard Nash-Sutcliffe efficiency (NSE), the results of the new viscous model for the Ventnor landslide in southern England, without an observation of cumulative error, show a better prediction ability (NSE = 0.98) than the traditional viscous model (NSE = 0.85). For the Utiku landslide of New Zealand, the NSE of the new viscous model achieves 0.87 compared to the traditional viscous model (NSE = 0.47) in Borehole 3; for Borehole PZA the NSE of the new and traditional viscous models are 0.61 and 0.18.

6.1 Introduction

Slope static instability means the process of reducing the safety factor (FS) to 1, or the formation of a slip surface because of global brittle damage. It manifests a deformation or tiny movement, not a big displacement along the slip surface (until the point B in Figure 6.1). For deep-seated static landslide problems, the current study focuses on the changed effective stress caused by groundwater fluctuation (not con-

6.1 Introduction

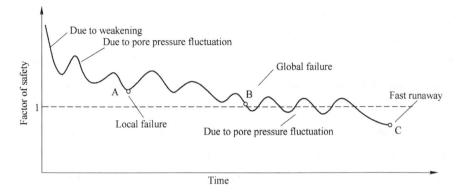

Figure 6.1 The physical process from a stable slope to a quasi static stage (the local and whole factor of safety is higher than 1 before A; between A and B the whole factor of safety is higher than 1 while the local factor of safety could be smaller than 1; After B, the whole factor of safety varies around 1) (modified from Picarelli, et al., 2004)

sider material strength changed) controlling the FS of slope. This is because the strength of the slip surface is usually treated as constant: an effective shear strength due to the slip surface locates under the water table (Fredlund et al., 1978; Fredlund and Rahardjo, 1993; Vanapalli et al., 1996; Duncan et al., 2014; Katte and Blight, 2015). In a quasi-static landslide analysis (movement after slip surface formation) (after the point B in Figure 6.1), the landslide evolution lasts a very long time, and could experience active, stop, reactive, acceleration, deceleration, and finally run away processes. Under these conditions, the strength of the slip surface treated as a constant is questionable. It has been recognized that besides changes in effective stress induced by pore pressure changed, shear strength of the materials changes caused by behaviour of the clay particles also governs landslide velocity (Lupini et al., 1981; Skempton, 1985; Angeli et al., 1996; Angeli et al., 2004; Picarelli, 2007). The cohesion in particular should be treated as a variable dependent on velocity, due to variable available water in the pores among soil particles. The reason for this is that the clay soil dilatancy and contraction caused by shear velocity causes porosity changes in the soil (Skempton, 1970; Houlsby, 1991; Manzari and Nour, 2000; Nakai and Hinokio, 2004; Dafalla, 2013). This variable porosity leads to changes of moisture and related strength, especially cohesion (Shimizu,

1982; Asaoka, et al. 1999; Nakai and Hinokio, 2004). The quasi-static landslide movement is therefore complexly determined by effective stress, water availability, and the development of shear strength (cohesion).

As one of the most comprehensive models to describe the quasi-static landslide, the viscous model is widely employed. An early viscous model was applied to a mudslide in Cortina d'Ampezzo, Italy, by Angeli et al. (1996). This model showed an agreement between the calculated displacement and the recorded displacement, by introducing the velocity dependent on viscous resistance. Following that, similar viscous models were used in many landslides, including the Alverà landslide located in the Italian Dolomites, near Cortina d'Ampezzo (Ranalli et al., 2010), the landslide (over-consolidated clay) at Cortina d'Ampezzo (Italy) (Gottardi and Butterfield, 2001), the Bindo-Cortenova translational landslide (Italian Prealps, Lombardy, Italy) (Secondi et al., 2013), the landslide near the Rhine River valley in the Vorarlberg Alps, Austria (Wienhöfer et al., 2009), the Vallcebre landslide (Eastern Pyrenees, Spain) (Corominas et al., 2005), the Super-Sauze landslide located in the south French Alps in the Barcelonnette Basin, on the left bank of the Ubaye River (Bernardie et al., 2014), and the Portalet landslide (Sallent de Gállego, Central Spanish Pyrenees) (Herrera et al., 2009). This traditional viscous model considers a local velocity calculation (object is slope unit (Figure 2.5)) and can be expressed by Equation. (2-12) and Equation. (2-13).

The common characteristics of this kind of viscous landslide model may produce three challenges.

(1) Based on the force mechanism of infinite slope model.

An infinite slope is usually defined as a long shallow landslide. In reality the infinite slope model is applied to a local, not global, movement and is also strictly limited by geometry (Pack, 2001; Borga et al., 2002; Casadei et al., 2003; Acharya et al., 2006; Claessens et al., 2007). If the geometry of the landslide does not match the infinite slope, the prediction of the landslide displacement rate would be challenged.

(2) Ignoring drive force changes caused by water availability.

Water availability affecting the drive force depends on the changes of slope weight caused by water infiltration.

(3) Treating the cohesion of material as a constant.

All the cases above treat the cohesion as a constant residual value. The potential strength variation deriving from soil dilatancy and contraction caused by variable velocity should be considered.

6.2 New viscous velocity based model

According to the three potential challenges above, the new viscous landslide model will improve the slope displacement prediction from three aspects:

(1) Since there is more application of the limited equilibrium model than the infinite slope model (Collison and Anderson, 1996; Wilkinson et al., 2002; van Beek and van Asch, 2004; Talebi et al., 2007), the slice method-based limit equilibrium model is used for movement mechanism, which has been applied in two deep-seated landslides of the Trièves Plateau (Van Asch et al., 2009).

(2) The slope weight is considered as a function of available water, not a constant, which makes the driving force of landslides variable.

(3) A cohesion-dependent on velocity module is developed, reflecting the soil consolidation and shear dilation action affecting the cohesion.

6.2.1 Introduction of new viscous model

Compared to the traditional viscous model, the shear velocity-strength module and water-drive force module regard quasi-static landslide movement controlled by the change of effective stress, change of slope mass weight, and development of shear strength (Figure 6.2(a)). Firstly, the initial material parameters including slip surface initial strength, density of mass, viscosity of slip surface, and initial porosity of the mass are input into this model, in addition to the geometry and groundwater table information. Secondly, the effective stress and velocity-strength modules are used to calculate the resistant force, while the water-drive force module is employed for calculation of driving force changed. The velocity-strength module has two functions: (1) During the fast landslide stage, as shown in Figure 6.2(b), the dilatancy of clay increases the porosity and water fills the pore, improving more moisture. A lubrication effect (bonds of particles are more easily broken) from the water decreases cohesion in the clay and damages the internal particle bond strengths. Thus, the

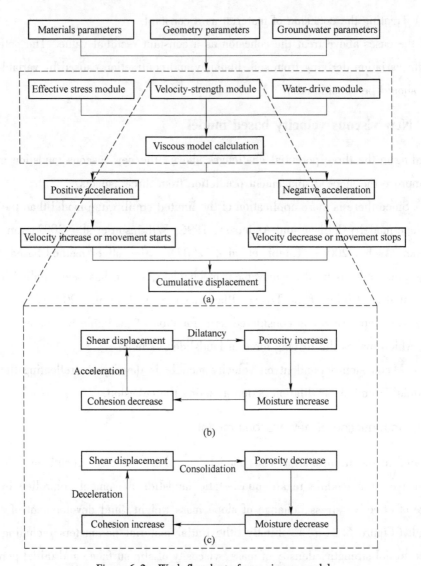

Figure 6.2 Work flowchart of new viscous model

(a) Work flow of whole model. The initial parameters include material, geometry, and groundwater inputs; the model considers variable effective stress, strength dependent on velocity, and variable mass density induced by change of groundwater table; (b) Acceleration stage of velocity-strength module; (c) Deceleration stage of velocity-strength module

strength of landslide mass becomes lower and the slope accelerates. Figure 6.3 indicates the dilatancy process of clay under undrained shear force. It shows that the volume of the shear zone expands under time domains and describes the development of cracks and the formation of particle fragments. (2) As shown in Figure 6.2(c), when a landslide moves very slowly or stops, the normal stress of slope may play a more obvious function in consolidation of the clay. The porosity of clay decreases and the water is squeezed out of the pores. The decrease of moisture makes the bonds of particles combine more closely. The consolidation process of clay soil is described in Figure 6.4. During the very slow landslide or stable process, the normal stress squeezes the soil and decreases the porosity for consolidation. This process could be a drained condition, and water is squeezed out of the pores. It should be emphasised that these two processes might take a very long time and the landslide would undergo the processes of movement, stop, and new movement during the long creep time (Auzet and Ambroise, 1996; Wienhofer et al., 2009; Li et al., 2010; Di Maio et al., 2013). Thirdly, the model output includes the cumulative displacement and velocity.

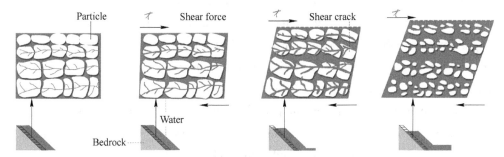

Figure 6.3 Dilatancy process of clay under undrained shear force. Increased pores in soil lead to more water get into the pores of soil. Thus, due to more water the bonds among soil particles are easily damaged under shear force, which decreases the cohesion of material (fast movement)

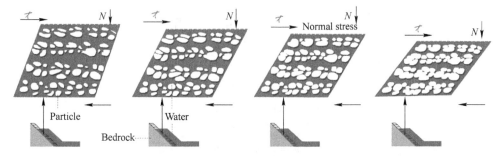

Figure 6.4 Shear consolidation process of clay soil. Increased normal stress squeezes the water out of the pores in soil. Thus, due to less water and pores, the bonds among soil particles are easily recovered which increases the cohesion of material (slow movement)

6.2.2 Velocity-strength module

For the velocity-strength module, physically, landslide movement first affects the porosity and moisture of the slip surface. Until now, the direct relationship between the displacement rate and the porosity has only been roughly investigated in the laboratory (Shimizu, 1982; Asaoka et al., 1999; Nakai and Hinokio, 2004) and by numerical cases (Xu et al., 2006). Most studies consider the porosity to change with dilatancy volume rate in the laboratory (Skempton, 1970; Bolton, 1986; Houlsby, 1991; Manzari and Nour, 2000; Nakai and Hinokio, 2004; Dafalla, 2013). Furthermore, laboratory tests indicate cohesion (strength component) changes with the soil moisture and porosity (undisturbed and remoulded samples) (Figure 6.5). In addition, an increase of moisture content reduces the cohesion by reducing the suction and surface tension. In Figure 6.5(a), the increase of consolidation time means a porosity of soil reduction under a compressive force. Consequently, the increase of the inter particle bond strength demonstrates the improvement of cohesion (Ochepo et al., 2012). The reduction in compaction decreases cohesion through the increasing porosity and essentially weakens the bond force as shown in Figure 6.5(b) (Gao et al., 2013). Similarly, the clay content reduction means there are fewer fine particles, and therefore, the increased porosity lowers the cohesion as shown in Figure 6.5(c) (McKyes et al., 1994; Dafalla, 2013). In Figure 6.5(d) the reduced density of the material produces higher po-

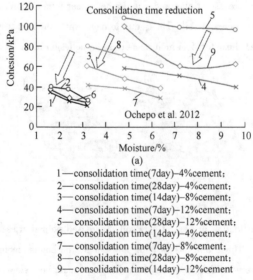

(a)

1—consolidation time(7day)–4%cement;
2—consolidation time(28day)–4%cement;
3—consolidation time(14day)–8%cement;
4—consolidation time(7day)–12%cement;
5—consolidation time(28day)–12%cement;
6—consolidation time(14day)–4%cement;
7—consolidation time(7day)–8%cement;
8—consolidation time(28day)–8%cement;
9—consolidation time(14day)–12%cement

6.2 New viscous velocity based model

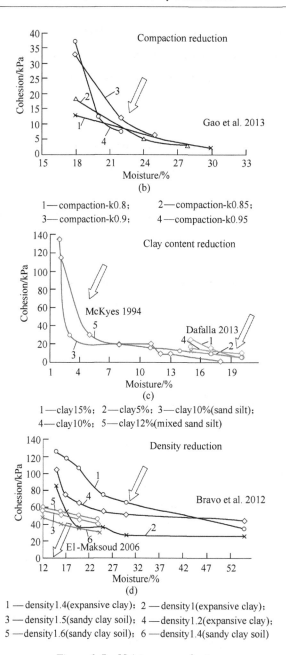

Figure 6.5 Moisture vs. cohesion

(a) Reducing consolidation time (consolidation time affecting the cohesion-moisture relationship) (Ochepo et al., 2012); (b) Reducing compaction (compaction degree affecting the cohesion-moisture relationship) (Gao et al., 2013); (c) Reducing clay content (clay content affecting the cohesion-moisture relationship) (McKyes et al., 1994; Dafalla, 2013); (d) Reducing density (density affecting the cohesion-moisture relationship) (El-Maksoud, 2006; Bravo et al., 2012) rosity and reduces the cohesion (El-Maksoud, 2006; Bravo et al., 2012). In summary, the key point of the velocity-strength module is to bridge the cohesion and velocity (displacement rate) with the porosity (moisture).

6.2.3 New viscous model calculation

The new viscous model considers the whole movement of landslide based on cumulative slices. The force calculation under one time unit is as follows: Figure 6.6 is the slices analysis and geometry of slope. Equation (6-1) calculates the resistance force of a slope, while Equation (6-2) calculates the drive force of slope based on cumulative slices. Equation (6-3) calculates variable unit weight of slope mass (water-drive force module). Equation (6-4) shows that the general viscous model consists of resistance force, shear force and viscous force based on Newton's second law. Equation (6-5) is the final expression to describe quasi-static landslides.

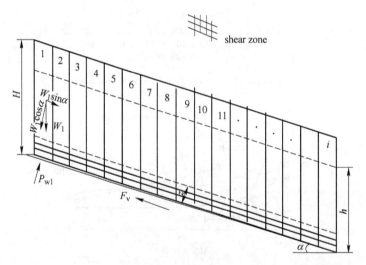

Figure 6.6 Geometry and force analysis in the new viscous landslide model. The model is based on the limit equilibrium method and can calculate the whole displacement of a complex geometry slope

W_1—weight of slice1; α—slope angle; F_v—vicous force; H—hight of slopemass; z—shear zone thickness; h—water table; P_{w1}—porewater pressure of slice1

$$T' = \sum_{i=1}^{n'} [c(v)w/\cos\alpha_i + (r_i w H_i \cos\alpha_i - P_{wi})\tan\phi] \qquad (6\text{-}1)$$

6.3 The Ventnor landslide, Isle of Wight, Southern England

$$F = \sum_{i=1}^{n'} (r_i w H_i \sin\alpha_i) \quad (6\text{-}2)$$

$$r_i = (H_i r_d + h_i r_w n)/H_i \quad (6\text{-}3)$$

$$F - T' = ma + F_v = ma + \eta \frac{v}{z} \quad (6\text{-}4)$$

$$\sum_1^i [(H_i r_d + h_w r_w n) w H_i \sin\alpha_i] - \sum_1^i [c(v) w/\cos\alpha_i$$

$$+ (H_i r_d + h_w r_w n) w \cos\alpha_i \tan\phi] = ma + \eta \frac{v}{z} \quad (6\text{-}5)$$

where, F and T' are slip force and resistance force of whole slope; $c(v)$ is the velocity-cohesion module; w is the width of slice; α_i is the slice slope angle; r_i is the unit weight of landslide; H_i is the height of slice; r_d is unit weight of dry landslide; h_i is the height of the ground water table; P_{wi} is the pore water pressure of slice base; r_w is unit weight of water; n is the porosity of the landslide; ϕ is the friction angle; m is the mass of whole slope; F_v is the viscous force; a is the acceleration; η is the viscosity; v is the velocity; z in our case is the height of the shear zone.

6.3 The Ventnor landslide, Isle of Wight, Southern England

6.3.1 Introduction to Ventnor landslide, Isle of Wight

The Ventnor landslide is a deep-seated landslide and poses a hazard to over 6000 residents on the south coast of the Isle of Wight, UK. Related previous studies incorporate slip mode, ground behaviour, climate, mechanisms and causes (e.g., Moore et al., 1995, 2007a, 2007b). The unstable area of the Ventnor landslide comprises 0.7km^2, and main scarp in the "Lowtherville Graben" reduces the current extent of instability (Figure 6.7(a)). Moore et al. (2007a) used building and geophysical investigations from 2002 and 2005 to portray of the materials of landslide types. Figure 6.7(b) demonstrates the details of material units (for more details about the geology structure and background see Carey (2011)). The whole landslide moves slowly (generally mm-cm/year). Heavy rainfall will increase the velocity of the landslide (Moore et al., 2010).

6.3.2 Monitoring data in the model (displacement and pore water pressure)

The main monitoring data includes pore water pressure from the vibrating wire pie-

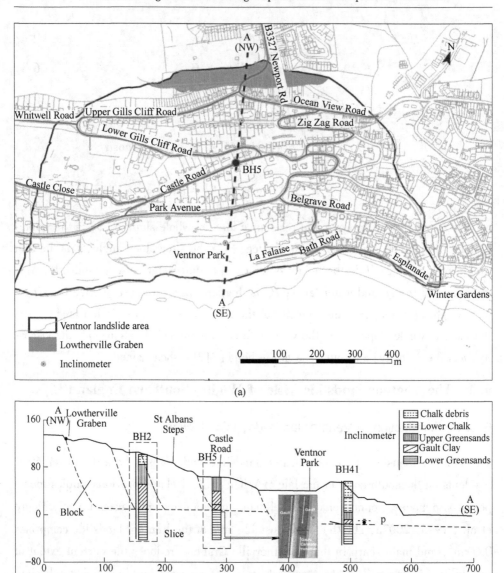

Figure 6.7 Schematic illustration of the Ventnor landslide (after Carey and Petley, 2014) (a) Map of urbanisations within the Ventnor landslide area; (b) Schematic illustration of the landslide cross-section indicating the low angulate shear surface, the landslide blocks, and the positions of detector devices (machine-driven piezometers at the Winter gardens (P); C is the crackmeters)

zometers P, and the continuous displacement from crackmeters C (Figure 6.7(b)). The piezometers recorded groundwater level and the crackmeters continually monitored the main movement between 1995 and 2002 at 24-hour intervals. A rela-

6.3.3 Strength properties of the main slip surface

Gault clay (main slip surface) was derived from the borehole BH5. A suite of laboratory tests has been used to determine the geotechnical characteristics of the Gault clay within the Ventnor landslide complex. A series of isotropic, consolidated, undrained (ICU) triaxial tests were applied to establish field-stress conditions for the peak and residual strength envelope (Figure 6.8). The natural moisture of Gault

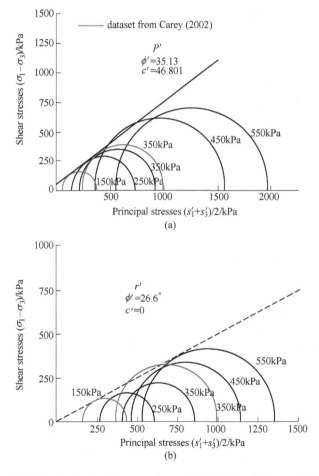

Figure 6.8 ICU Mohr Coulomb failure envelopes of Gault Clay (Carey and Petley, 2014)
(a) Peak strength envelope of undisturbed soil samples; (b) Residual strength envelope of undisturbed soil samples. Consolidation can recover the cohesion of materials

Clay samples is around 17% and plastic limit and liquid limit are 21% and 56%, respectively. The volume (moisture) of clay sample increases over 5% when initial soil transforms into residual status according to Carey and Petley (2014). That means the residual soil exists as a viscoplastic status since the moisture is between plastic limit and liquid limit. Thus, the viscous model is properly suitable model for describing the quasi-static landslide.

6.3.4 Application of a new viscous model at the Ventnor landslide

6.3.4.1 Basic assumptions of the model

(1) Ignoring the interaction force between neighbour slices like Fellenious method (Figure 6.9).

(2) All the slices are treated as a whole mass movement (Figure 6.9).

(3) The horizontal movement of C (main crack change rate) in Lowtherville Graben is treated as the movement of landside.

(4) Change of the groundwater table (data of P) is regarded the same under the same time unit in the vertical direction.

(5) Friction uses the residual value, and cohesion depends on the velocity (variable between residual and peak values).

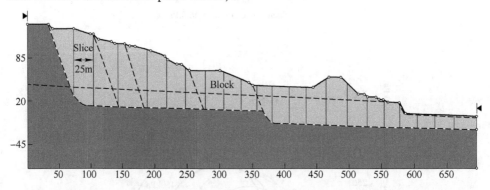

Figure 6.9 Force and slices analysis of geometry model of Ventnor landslide
(based on the Figure 6.7(b))
(Note: the first and second slices in the landslide head consider the component of forces along the main slip surface)

6.3.4.2 Procedures

(1) Determination of the geometry model and the initial strength parameters. Force

analysis of the geometry model is based on Figure 6.9 (average dry density and average porosity are 1900kg/m³ and 0.2, respectively) and the initial friction angle and cohesion are 16° and 45kPa, respectively. Theinitial cohesion considers the consolidation, recovers the bonds of particles and increases the strength. The friction angle is lower than in the laboratory test (26.6°) which makes the slope mobile. The height of the shear zone is assumed to be 1m. According to Moore et al. (2006), the initial constant groundwater table is around 53m above the bedrock.

(2) With the current model framework the pre-viscous coefficient is determined using a back analysis, considering a constant cohesion (average viscous coefficient of the first five days) (01.01.1998~05.01.1998); then by increment or decrement of the pre-viscous coefficient, 300 days of measurement of displacement (01.01.1998 ~01.10.1998) is used to calibrate the viscous coefficient to make a minimum convergence error (calibration in Figure 6.11). For example, the initial viscosity is 5.2×10^{11} and is reduced every time step by 0.1×10^{11} until the minimum error of 4.7×10^{11} is achieved.

The final viscous coefficient (4.7×10^{11} Nsm^{-3}), mainly derived from material property, is close to the direct shear value (4.5×10^{11} Nsm^{-3}) obtained in the laboratory but larger as the final value (7.8×10^{7} Nsm^{-3}) used in the model of Angeli et al. (1996). It should be pointed out that the calculation of the viscous coefficient and shear zone height are treated as a whole objective. This means the assumed shear zone height would affect the viscous coefficient value. In addition, the 300 measurement days are used for calibration and the rest data set is used for validation.

(3) The measurement data is used to back analyse the coefficients of c-v relation (velocity-strength module) for the data fitting of velocity and cohesion as shown in Figure 6.10. The function of c-v is incorporated into the model as input data of a velocity-strength module.

(4) Input of monitoring data into the model and simulation generation.

6.3.5 Prediction results and analysis

In Figure 6.11 (a), cv-linear model (new viscous model) considers the simple linear relationship between cohesion and velocity. c-constant model (original viscous model) reflects a constant cohesion. Part 1 shows both the considerations made for

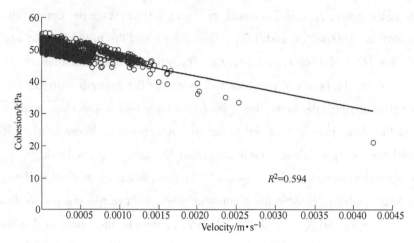

Figure 6.10 Cohesion vs. velocity based on real back analysis of Ventnor landslide. The data belt shows a obvious linear relationship between landslide velocity and cohesion ($R^2 = 0.594$)

the velocity increase and decrease when the groundwater table moves up and down and creates no obvious differences. In Part 2 the trend of both models is similar to the observation data although both the models show underestimation of the cumulative displacement. The stronger evidence supporting the cv-linear model is from Part 3. The reduced resistant force accelerates the movement of the slope under low effective stress. Consideration of normal stress alone does not adequately describe the acceleration process unless the strength reduction is caused by an increased velocity. In other words, only considering the pore water pressure makes it difficult to copy the velocity changes. Especially in the final runaway stage of the landslide, it would possibly be unnecessary for a "huge pore water pressure" threshold to trigger a landslide disaster with the strength-velocity module. Next, velocity calculation is shown as in Figure 6.11(b). The cv-linear model agrees with the real monitoring data much better compared to the c-constant model.

6.4 The Utiku landslide in New Zealand

6.4.1 Introduction to Utiku landslide, New Zealand

The Utiku landslide is a large deep-seated, translational landslide which has a low movement velocity. One gentle fault from the east-west compression produced the Utiku landslide regional tectonic setting (Lee et al., 2012). Related uplift drives

6.4 The Utiku landslide in New Zealand

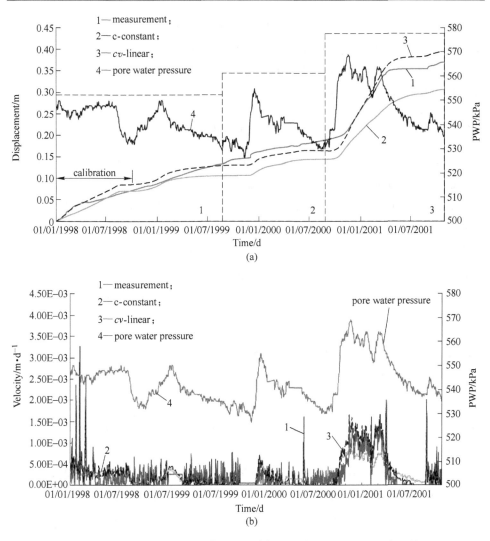

Figure 6.11 Results of the prediction model using data from Jan. 1998 to Nov. 2001 of the Ventnor landslide

(a) Cumulative displacement. cv-linear model can describe the real displacement better than c-constant model and does not have an obvious cumulative error compared to the c-constant model; (b) Velocity analysis from new and traditional models. cv-linear model can describe the measurement velocity much better than the c-constant model especially during the periods of high velocity

fluvial incision with rates about 1.5mm/y to 2.0mm/y (Pillans, 1986; Pulford and Stern, 2004).

The main geological materials include landslide debris (intact blocks of sandstone, partially remoulded rafts of intact sandstone and remoulded sandstone), landslide slip surface clay, river-terrace gravels, the in-situ Tarare sandstone; and the in-situ Taihape mudstone. The assumed slip surface is a thin clay layer (thickness from 0.05m to 0.2m).

6.4.2 Involved monitoring and strength date

From July 2008 on, four continuous GPS measured the surface movements. Full details of the monitoring system are contained in Massey (2010). Pore pressures within the landslide were measured using piezometers in boreholes. Hourly readings from vibrating-wire piezometers were converted to pressure and averaged over each 24-hour period (accuracy is ±0.1%). Geotechnical tests of the clay from the slip surface were applied using a ring-shear apparatus (Kilsby, 2007; Massey, 2010). The residual strength of the slip surface was a cohesion of (4±6) kPa and a friction angle of 8.3°±1°.

6.4.3 Application of new viscous model to Utiku landslide

6.4.3.1 Basic assumptions of the model

(1) Ignoring the interaction force between neighbour slices like Fellenious method (Figure 6.12(b)).

(2) All the slices are treated as a whole mass movement (Figure 6.12(b)).

(3) The horizontal movement of GPS (surface movement) in PZA and BH3 is treated as the movement of the landside.

(4) Changes to the groundwater table (data of P) is regarded as the same under the same time unit in the vertical direction.

(5) Friction uses the residual value, and cohesion depends on the velocity (variable between residual and assumed peak values).

6.4.3.2 Procedures

(1) Determination of the geometry model and the initial strength parameters. Force analysis of the geometry model is based on Figure 6.12(a) (average dry density and average porosity are 1900kg/m^3 and 0.2, respectively) and the initial friction angle and cohesion are 7° and 8kPa, respectively. The initial cohesion considers that the

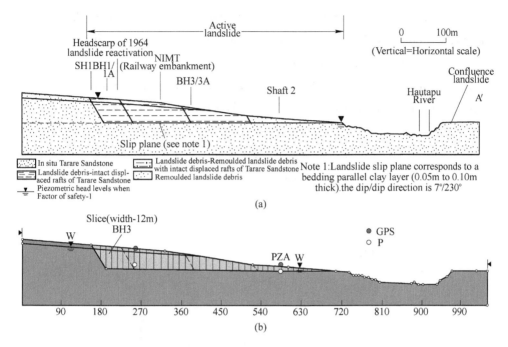

Figure 6.12 Utiku landslide of New Zealand

(a) Geological profile (Massey et al., 2013);

(b) Force and slices analysis of geometry model (based on Figure 6.12(a))

(Note: the first and second slices in the landslide head consider the component of forces along the main slip surface)

consolidation recovers the bonds of particles and increases the strength from 4kPa to 8kPa. The height of the shear zone is assumed to be 0.1m. The initial constant groundwater table is according to Massey et al. (2013).

(2) For BH3, using the current model framework we determinate the pre-viscous coefficient by back analysis, considering the constant cohesion (average viscous coefficient of the first five days) (08.05.2009~12.05.2009); then by increment or decrement of the pre-viscous coefficient, 180 days of measurement of displacement (08.05.2009~08.11.2009) is used to calibrate the viscous coefficient to make a minimum convergence error (calibration in Figure 6.14(a)). For example, the initial viscosity is 5.2×10^{11}, every time we reduce the 0.1×10^{11} until the 4.7×10^{11} which produces the minimum error with recorded data; The final viscous coefficient (4.8×10^{11} Nsm^{-3}), mainly decided by material property, is close to the value in laboratory (4.5×10^{11} Nsm^{-3}) but over to the final value (7.8×10^{7} Nsm^{-3}) used in the model of Angeli et al. (1996). It should be pointed out that the calculation of the

viscous coefficient and shear zone height is treated as a whole. It means the assumed shear zone height would affect the viscous coefficient value. In addition, the 180 days' measurement is for calibration and the rest 620 days' data is for validation.

(3) The measurement data is used to back analyse the coefficients of the c-v relationship (velocity-strength module) for the data fitting of velocity and cohesion as shown in Figure 6.13. The function of c-v is then input into the model as a velocity-strength module. The data distribution (the same velocity correlates to the different cohesions) is because the same velocity value is produced by different stress condi-

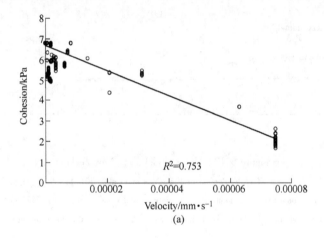

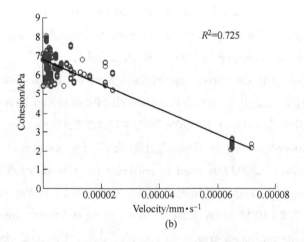

Figure 6.13 Cohesion vs. velocity based on real back analysis of the Utiku landslide
(a) BH3. An obvious linear relationship is observed ($R^2 = 0.753$);
(b) PZA. An obvious linear relationship is observed ($R^2 = 0.725$)

tions (negative, positive, or constant acceleration). These complicated stress conditions lead to variable deformations, porosities of slip surface furthermore cohesion even under the same velocity.

(4) Input the monitoring data into the model and make a simulation.

(5) For PZA, all the parameters take the same as in BH3 since this landslide is treated as a whole movement.

6.4.4 Prediction results and analysis

In Figure 6.14, the linear relationship as a velocity-cohesion module is coupled into the viscous model. c-constant model reflects constant cohesion. In Figure 6.14(a), pore pressure is identical during Part 1 and Part 2, the changed velocity induced by reducing groundwater is lower than changed velocity during increasing groundwater (Bertini et al., 1984; Picarelli, 2007; Gonzalez et al., 2008). The same phenomenon happens in Part 1 and Part 2 of Figure 6.14(b). The main reason could be a constant viscosity in the viscous model results in a linear relationship between pore pressure and displacement rate (Gonzalez et al., 2008) while the laboratory test indicates viscosity inversely proportional to velocity (Massey, 2010). That shows potential higher velocity at the acceleration stage. In summary, this new viscous model can simulate the whole landslide process well. Especially, the stronger evidence supporting the cv-linear model is from the Part 3. The reduced resistant force accelerates the movement of slope under low effective stress. However, pure consideration of normal stress does not adequately describe the acceleration process unless the strength reduction is caused by increased velocity. In other words, only considering the pore water pressure is difficult to reproduce the velocity changes. In Figure 6.14 (c) and Figure 6.14(d), the new model improves the simulation of movement rate in the same time step.

6.5 Discussion

6.5.1 Physical interpretation

In Figure 6.1, a local failure (local brittle damage) takes place and expands into the stable slope during times of weakening or pore water pressure fluctuation. Until general brittle failure (formation of a viscous shear zone), the landslide is a quasi-

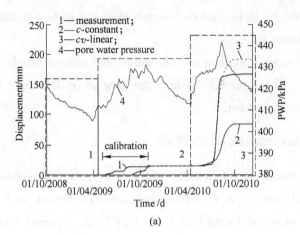

(a)

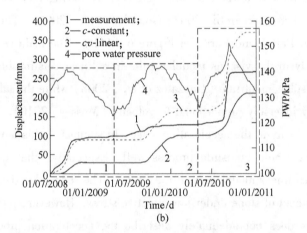

(b)

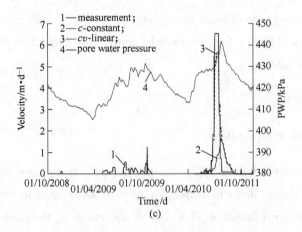

(c)

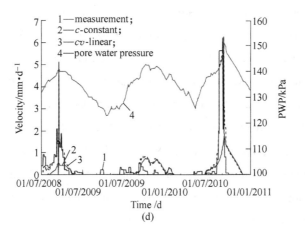

Figure 6.14 Results of the prediction model. The cv-linear model describes the displacement better than the c-constant model during high PWP period while velocity calculation of the cv-linear model is always close to real velocity

(a) Displacement of BH3 (data from Oct. 2008 to Dec. 2010); (b) Displacement of PZA (data from Jul. 2008 to Jan. 2011) of the Utiku landslide; (c) Velocity of BH3 (data from Oct. 2008 to Dec. 2010); (d) Velocity of PZA (data from Jul. 2008 to Jan. 2011) of the Utiku landslide

static movement (post-failure) (Picarelli et al., 2004; Petley et al., 2005). During the process the friction is a function of asperities on the sliding plane that are worn off with increasing displacement (Lucas et al., 2014). This was thus considered a residual value after point B in Figure 6.1. Cohesion is a function of inter-particle bond strength. In many materials cohesion decreases as soon as bonds are broken (i. e., dry, intact rock), and when clay is present the plasticity describes the change in cohesion with water content. Bonds can be broken and reform, and their strength is mainly determined by water availability. Thus, the soil cohesion recovers due to consolidation at the post-failure stage. During the post-failure stage, quasi-static analysis or pseudo-dynamic (slow movement) creep is also controlled by cohesion at the shearing plane of the landslide. On micro scale, in the new viscous model, the velocity-strength module makes the modelling more physically. Some macroscopic field evidence or observations also support the phenomenon of velocity-dependent strength. Consolidation affecting the shear strength in viscous landslide modelling was first revealed by Angeli et al. (1996) and is the main reason for error estimation. With the increase of consolidation (very slow velocity period), the landslide needs higher pore water pressure to restart. Thus, cohesion in a landslide may not actually be is a constant. Some other evidence suggests that continuous, long-term landslide motion may be partly the result of the dilation and consolidation of fine-grained landslide material (Keefer and Johnson, 1983; Baum and Johnson, 1993; van Asch et al., 2007). Decelerating, short-distance landslide movement has been ascribed to increases of effective stress and shear strength resistance along the landslide base (e.g., Iverson et al., 1997, 2000; Moore and Iverson, 2002; Iverson, 2005). Similarly, semi-continuous landslide motion over long periods (months-years) may result from dilation during shear displacement. Subsequent consolidation allows recurrence of shear-induced dilation when pore-pressures rise sufficiently to trigger renewed movement. This behaviour has been observed at laboratory (Moore and Iverson, 2002) and field scale (e.g., Iverson et al., 2000) such as the continuously moving Slumgullion landslide in Colorado, USA (Schulz et al., 2009), and in a theoretical study (Schaeffer and Iverson, 2008). More importantly, as noted by Moore and Iverson (2002) and Iverson (2005), sufficient landslide displacement will cause dilations to steady-state porosity and deceleration associated with consolidation will no longer occur. Runaway landslide acceleration or landslide disaster may follow. In other words, if the strength reduction by shear dilation is ignored and the velocity is controlled by effective stress alone, the trigger threshold for a landslide disaster could be underestimated.

6.5.2 Parameters interpretation

In the calibration process, every initial cohesion produces an optimal viscosity (and optimal coefficients of c-v linear relation followed). These viscosity and coefficients could be different depending on the input of different initial cohesions. This is a common character of calibration-based model. But the model can still work well because these parameters can balance each other during the calibration process (for instance, a high initial cohesion results in a high viscosity; a low initial cohesion makes a low viscosity). In the UK case, the initial cohesion is 47kPa (peak value) when the pore water pressure is 550kPa (01.01.1998). However, in fact the peak cohesion could appear on 01.11.1999 (PWP is 530kPa). Thus, the initial cohesion for viscosity-calibration is a little bit "high" which produces an overestimation of the viscosity. The similar situation in the New Zealand case is that the initial cohesion is 8kPa (residual value) when the pore water pressure is 405kPa (lowest value during the monitoring period). In fact, the cohesion could be expected to increase to a higher value because of consolidation in the current time domain. Thus, the initial cohesion for viscosity-calibration is a little bit "low" which produces an underestimation of viscosity.

6.5.3 Comparison between traditional and new viscous model

6.5.3.1 Pore water pressure and velocity

For the same pore water pressure, the variable velocity of the landslide could depend on the trend of changed pore water pressure, that is, when the pore water pressure is rising or falling (Bertini et al., 1984; Van Asch et al., 2007; Gonzalez et al., 2008; Matsuura et al., 2008). Thus, the non-linear relation between pore water pressure and velocity is implied. According to Gonzalez et al. (2008), a constant viscosity produces a linear relationship between pore pressure and velocity. Whether in the traditional or the new viscosity model, the constant viscosity setting could introduce the velocity error induced by same pore water pressure in different rising or falling time. In fact, the current velocity depends on both current acceleration (mainly decided by pore water pressure) and the velocity in the previous time step. Plus, it mentioned that the viscosity might depend on velocity. All these considerations make the model not easily reproduce the velocity triggered by rising or falling pore pressure.

6.5.3.2 Cohesion depended on velocity

The traditional viscous model considers the strength as a constant. By contrast, the strength (cohesion) is treated as a variable in this new model. Figure 6.10 is the back analysis of the relationship between cohesion and velocity in the UK case. Figure 6.13 is the back analysis of the relationship between cohesion and velocity of the New Zealand case. Both figures demonstrate that a linear function can generally describe the relationship between cohesion and velocity. Better description of c-v relation might depend on the fitting method, suitable function, and even advanced algorithm. In addition, when the velocity is zero, the cohesion varies in a wide range. It could conclude that the consolidation time and consolidation pressure are also important factors of affecting the cohesion recover.

6.5.3.3 Strength description during the whole process

The whole process could be considered in four stages, as shown in Figure 6.15.

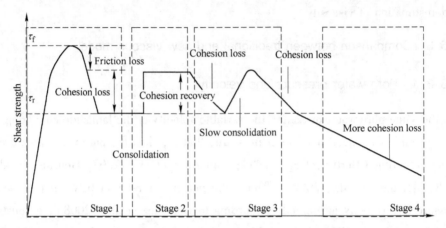

Figure 6.15 Conceptual strength changes in quasi-static landslide movement. The strength of material changes from the peak value to the residual value. Material losses the friction while the cohesion (loss or recovery) depends on the bonds among soil particles

Stage 1: is the situation before landslide movement occur (e.g. Wang et al., 2010; Schulz and Wang, 2014; Kimura et al., 2014; Carey and Petley, 2014). Carey and Petley (2014) indicated the volume increase during the clay shear test, which meant water content improvement. The cohesion began to reduce, but the loss of cohesion is not enough to produce this displacement, thus, the friction angle must drop until the residual value similarly to the idea of Skempton (1985), from peak to residual strength, this is partly due to moisture increase and partly from the result of particle rearrangement.

Stage 2: If the groundwater does not decrease, the landslide would keep moving until failure. If there is a groundwater reduction, the landslide would stop and begin to consolidate. The water in pores would flow out because of the increased normal stress. This cohesion recovery is lower (or equal to) than the cohesion loss in stage 1. It is called a new "residual strength".

Stage 3: If the groundwater drops slowly, the landslide would decelerate and begin to slowly consolidate. The recovery of cohesion needs more time.

Stage 4: When groundwater increases, the landslide begins to move (shear dilatancy producing more porosity). If there is no decreased groundwater table, the cohesion would continuously reduce and the landslide would accelerate until the disaster occurs.

There are some questions to be asked, however.

In Stage 2, many studies determine the result from Gibo et al. (2002): when the normal stress is lower than 100kPa, the strength recovery is obvious; when the normal stress higher than 100kPa, it is not obvious (Carrubba and Del Fabbro, 2008; Stark and Hussain 2010; Mesri and Huvaj-Sarihan, 2012). Their research used the same consolidation stress to recover the strength (deviator stress unchanged); while the condition used in this study was a groundwater drop inducing the normal stress increase (deviator stress changed). Some of these studies (Carrubba and Del Fabbro, 2008; Stark and Hussain, 2010; Mesri and Huvaj-Sarihan, 2012) use unloading shear stress to loading shear stress action during the residual stage for testing the recovered strength. They found the obvious residual strength increase (it is a kind of changed deviator stress). Changed deviator stress could recover the strength to a great extent. A good case is that of Miao et al. (2014). When normal stress increased to 150kPa (shear stress unchanged), the "heal" from the changed deviator stress is obvious.

Stage 3 considered cohesion loss vs. velocity. In fact, there are few studies on this topic. All the residual strength vs. shear rate (Wang et al., 2010; Schulz and Wang, 2014; Kimura et al., 2014) should be at stage 1. Lemos et al. (1985) found that the slow shearing caused a slowly changed strength, and a fast shearing produced a first increased then decreased strength during the residual stage. Tika and Hutchinson (1999) found a fast shearing greatly reduced the strength during the residual stage.

The clay in this study may be unsaturated because of high normal stress (water can not easily get in and air can not easily get out). The cohesion-velocity relationship in the model can be supported by Matsushi and Matsukura (2006) and Rahardjo et al. (1995). It involved a changed cohesion and unchanged friction angle, cohesion (water content) vs. normal stress.

6.5.3.4 Prediction ability

The first application of the viscous model of the Cortina d'Ampezzo mudslide by Angeli et al. (1996) shows a high accuracy. In a four year monitoring period, the prediction deviation of cumulative displacement is no more than 20cm and there is no cumulative error as time elapses (i.e, the prediction deviation does not increase). However, since the calculated value is lower than the recorded value in the dry season, a cohesion recovery is possible due to consolidation. In the six year monitoring periods of the Super-Sauze landslide (Bernardie et al., 2014) the prediction error of traditional viscous model increased from around 8cm to around 70cm. This is a typical natural limitation of the no-self-adjusting model. The cumulative error will continuously increase if there is no model refreshment. Similarly, the Alverà landslide displacement prediction based on a traditional viscous model (Ranalli et al., 2010) showed low errors in eight years (no more than 5cm and no cumulative error) due to four new calibrations. The Heumöser landslide displacement prediction in the Vorarlberg Alps, Austria (Wienhöfer et al., 2009), suggests that consideration of the variable slope density of water infiltration is a more accurate prediction (no more than 9cm deviation). Unfortunately, as there is not more than one year monitoring data, an increase of the cumulative error cannot be demonstrated. The application of a viscous model in the Vallcebre landslide (Eastern Pyrenees, Spain) (Corominas et al., 2005) shows good prediction ability in borehole S-2 and S-9 in almost 1.5 years with 2~4cm deviation. The main material of the slip surface in S-2 and S-9, however, is fissured shale (more brittle not cohesive) considering the cohesion-velocity module does not work here. The prediction of clayed siltstone (cohesive material) movement in borehole S-11 produces a bigger error, achieving 20cm without a cohesion-velocity module. The geometry problem (infinite model is not suitable here) by Corominas et al. (2005) could also be considered. By contrast, in the UK case, the biggest error of our new viscous model in the four-year monitoring period is around 5cm without cumulative error, and in the New Zealand case, for both PZA and BH3, the cumulative errors are no more than 5cm in three years. In order to evaluate the performance of the new viscous model, the book introduces the standard Nash-Sutcliffe efficiency (NSE, defined by Nash and Sutcliffe, 1970) which is the criteria most widely used for the calibration and evaluation of models with observed data (other criteria to assess models see Moriasi et al. (2007)), whereas NSE is dimensionless, being scaled onto the interval [−inf to 1.0]. NSE is taken to be the "mean of the observations" (Murphy, 1988) (i.e., if NSE<0, the model is no better than using the observed mean as a predictor). The equation of NSE is as follows:

$$\text{NSE} = 1 - \left[\frac{\sum_{i=1}^{n'} (Y_i^{obs} - Y_i^{sim})^2}{\sum_{i=1}^{n'} (Y_i^{obs} - Y_i^{mean})^2} \right] \quad (6\text{-}6)$$

where, Y_i^{obs} is the ith observation for the constituent being evaluated; Y_i^{sim} is the ith simulated value for the constituent being evaluated; Y_i^{mean} is the mean of observed data for the constituent being evaluated; n' is the total number of observations.

The NSEs of the traditional model and new model in the Ventnor landslide case are 0.85 and 0.98, respectively. This meansthatthe traditional model is not bad for describing the displacement, however, the modified new model is a better model, considering that the NSE of the best model is 1. In contrast, in the case of the Utiku landslide of New Zealand for the PZA, the NSE of the new model is 0.61 which hugely improves the prediction ability, considering that the NSE of the traditional model is 0.18. For BH3, the new model boosts the prediction ability (NSE = 0.87) compared to the traditional model (NSE = 0.47).

6.5.3.5 General characteristics

The traditional viscous model is based on an infinite slope model which has a relatively narrow application, and does not consider the effect of water availability on the driving force, which could produce an error especially in a steep deep-seated landslide. The strength (cohesion) is also treated as a constant in that model, while the new viscous model is based on the limited equilibrium method which is more widely used, and the model considers the density changes affecting the driving force. More importantly, it is an adaptive model which can reduce the cumulative error through a velocity-strength module based on the cohesion increment learning method.

6.5.4 Error analysis and outlooks

For the UK case, the new viscous model can relatively simulate the movement. While for the NZ land case, the main errors come from the velocity induced by same pore pressure during its rising or falling stage. Overall, landslide movement is difficult to trace. The potential errors could come from the differences between surface movements and slip surface movements, velocity inducing variable viscosity, and complex landslide geology and hydrogeology. For the viscous model, the improvement in the future could be considering the consolidation time or pressure, a better method for c-v relation, and a velocity inducing variable viscosity. For a landslide forecasting system, the main direction of research involves combining the groundwater prediction model including a deterministic model such as the Green and Ampt model, Richards equation, Van Genuchten equation, and Fredlund and Xing method (Fredlund and Xing, 1994; Chen and Young, 2006; Schaap and Van Genuchten, 2006; Weill et al., 2009) and calibration models such as the Tank model, and HBV model (Faris and Fathani, 2013; Abebe et al., 2010) with the new viscous model. The extended forecasting system can directly create a link between precipitation and landslide movement. In this forecasting system, the prediction and cumulative errors could be greater because of the greater uncertainty during the process of rainfall (snow accumulation and snow melt) infiltration, groundwater changes, and displacement variation.

6.6 Conclusions

Valuable conclusions include:

(1) In quasi-static landslide calculation, consideration of strength dependent on velocity is necessary for an improvement of prediction and physical meaning.

(2) The new viscous model based on water infiltration drive and limit equilibrium method can be applied more widely.

(3) The cohesion of slip surface and landslide velocity could have a linear relationship.

7 Conclusion

7.1 Key findings

Under the forecasting framework of groundwater induced deep-seated landslide, this study has combined precipitation with groundwater changes and the results of quasi-static landslide movement based on historic monitoring data. This study clarified the forecasting mechanism of precipitation induced groundwater and optimised the groundwater prediction model. In the quasi-static landslide phase, by considering a residual strength dependent on velocity and variable driving force due to water infiltration, the new viscous model can better describe the quasi-static landslide. Thus, the relationship between precipitation, pore water pressure, and movement measurements could generally be proved by these models above for the sophisticated forecasting systems.

(1) In deep-seated landslides, snow accumulation/melt and long infiltration paths mainly produce the time lag of groundwater. One important point is the time lag disturbs the accurate estimation of groundwater. Another important point is, the time lag could be an important component of early warning time (For example, by providing evacuation time from recognising the trigger event such as an extreme rainfall storm to the landslide disaster caused by the groundwater table). In this study, since there is maximum correlation between rainfall/snowmelt event and daily change of pore water pressure (daily changes of pore water pressure has higher correlation with rainfall/snowmelt events than the absolute pore water pressure), the daily change in pore water pressure is suggested as an index in groundwater prediction models. The modified tank model considers current changes of groundwater affected by the water infiltration of the current day and previous days. In order to eliminate the time lag error an equivalent infiltration method consisting of the proportion of infiltration of the current day and previous days was coupled with tank model. The straightforward method improves the application of the tank model in deep-seated landslides. In ad-

dition, coupling the snow accumulation/melt model makes this tank model usable in high latitudes zone. As an important step in hydro-induced landslides, this method links weather information (precipitation) and the groundwater table.

(2) The new landslide model uses slice method modelling rather than the infinite slope model since the geometry limitation of the infinite slope model narrows its application. Another aspect of this landslide model is consideration of available water changing the weight of slope mass and therefore the driving and resistant forces. An increase or decrease of the shear displacement rate affects the residual strength. Traditional viscous models do not consider the residual strength dependent on velocity, therefore, under-or-over-estimating the landslide velocity. In this landslide model, landslide velocity is subject to viscous force, the groundwater table, and residual strength affected by velocity of previous step. This model with the higher physical mechanism improves the accuracy of prediction.

7.2 Limitations and outlook

This modified tank model belonging to probability model still has not a strong physical meaning, and the calculation of snow accumulation may be not accurate enough because it depends on a simple statistical method. As with the other landslide models, initial calibration of new viscous landslide model is necessary to determine all parameters. Improvement of prediction ability is subject to initial continuous monitoring data feedback.

In the future, the groundwater produced by infiltration in heterogeneous or complex landslides needs validation. The definition of warning thresholds, including height of groundwater and landslide velocity, is also an important direction. The extension of early warning systems involving communication, message publication, and evacuation management should be considered.

References

Abebe N A, Ogden F L, Pradhan N R. (2010). Sensitivity and uncertainty analysis of the conceptual HBV rainfall-runoff model. Implications for parameter estimation. Journal of Hydrology, 389(3), 301-310.

Acharya G, De Smedt F, Long N T. (2006). Assessing landslide hazard in GIS: A case study from Rasuwa, Nepal. Bulletin of Engineering Geology and the Environment, 65(1), 99-107.

Agliardi F, Crosta G B, Zanchi A, et al (2009). Onset and timing of deep-seated gravitational slope deformations in the eastern Alps, Italy. Geomorphology, 103(1), 113-129.

Ahrens C D (2007). Meteorology today: An introduction to weather, climate, and the environment. Thomson Brooks. Cole, Canada.

Albers S C, McGinley J A, Birkenheuer D L, et al(1996). The local analysis and prediction system (LAPS): analyses of clouds, precipitation, and temperature. Weather and Forecasting, 11(3), 273-287.

Angeli M G, Gasparetto P, Silano S, et al(1988). An automatic recording system to detect critical stability conditions in slopes. Proceedings of the 5th International Symposium on Landslides, Lausanne.

Angeli M G, Gasparetto P, Menotti R M, et al (1996). A visco-plastic model for slope analysis applied to a mudslide in Cortina d'Ampezzo, Italy. Quarterly Journal of Engineering Geology and Hydrogeology, 29(3), 233-240.

Angeli M G, Pasuto A. (1998). A combined hillslope hydrology/stability model for low-gradient clay slopes in the Italian Dolomites. Engineering Geology, 49, 1-13.

Angeli M G, Pasuto A, Silvano S. (1999). Towards the definition of slope instability behaviour in the Alvera mudslide (Cortina d'Ampezzo, Italy). Geomorphology, 30(1-2), 201-211.

Angeli M G, Gasparetto P, Bromhead E. (2004). Strength-regain mechanisms in intermittently moving landslides. Proceedings of the 9th International Symposium on Landslides. London.

Asaoka A, Nakano M, Noda T, et al(1999). Progressive failure of heavily overconsolidated clays. Journal of the Japanese Geotechnical Society Soils & Foundation, 39(2), 105-117.

Auzet A. V, Ambroise B. (1996). Soil creep dynamics, soil moisture and temperature conditions on a forested slope in the granitic Vosges mountains, France. Earth Surface Processes and Landforms, 21(6), 531-542.

Avo-Almwirtschaftlicher verein oberbayern n. d. : Almbuch "Aggen Alm"- Ursprünglich amtliche Almdokumentation. Miesbach.

Baum R L, Johnson A M. (1993). Steady movement of landslides in fine-grained soils: A model for sliding over an irregular slip surface. US Department of the Interior, US Geological Survey.

Bernriede J. (1991). Unser Audorf, Chronik II. Teil, – 1014pp., Oberaudorf (Meißner-Druck

GmbH).

Bertini T, Cugusi F, D'Elia B, et al (1984). Climatic conditions and slow movements of colluvial covers in central Italy. Proceedings of the 4th International Symposium on Landslides, Toronto.

Bergström S. (1995). The HBV model. computer models of watershed hydrology. Water Resources Publications, Colorado, USA.

Bergström S, Carlsson B, Gardelin, M, et al (2001). Climate change impacts on runoff in Sweden assessments by global climate models, dynamical downscaling and hydrological modelling. Climate Research, 16(2), 101-112.

Bernardie S, Desramaut N, Malet J P, et al (2014). Prediction of changes in landslide rates induced by rainfall. Landslides, 12(3), 481-494.

Bernrieder J. (1991). Unser Audorf: Chronik II. Teil. Gemeinde.

Beven K, Germann P F. (1982). Macropores and water flows in soils. Water Resources Research, 18(5), 1311-1325.

Bloschl G, Kirnbauer R, Gutknecht D. (1991). Distributed snowmelt simulations in an alpine catchment I. Model evaluation on the basis of snow cover patterns. Water Resources Research, 27(12), 3171-3179.

Bocchieri J R. (1980). The objective use of upper air soundings to specify precipitation type. Monthly Weather Review, 108(5), 596-603.

Bodhinayake W, Si B C, Noborio K. (2004). Determination of hydraulic properties in sloping landscapes from tension and double-ring infiltrometers. Vadose Zone Journal, 3(3), 964-970.

Bolton M D. (1986). The strength and dilatancy of sands. Géotechnique, 36(1), 65-78.

Bonnard C, Noverraz F. (2001). Influence of climate change on large landslides: Assessment of long term movements and trends. Proceedings of the International Conference on Landslides Causes Impact and Countermeasures. Essen.

Borga M, Dalla Fontana G, Cazorzi F. (2002). Analysis of topographic and climatic control on rainfall-triggered shallow landsliding using a quasi-dynamic wetness index. Journal of Hydrology, 268(1-4), 56-71.

Bourgouin P. (2000). A method to determine precipitation types. Weather and Forecasting, 15, 583-592.

Brammer D D, McDonnell J J. (1996). An evolving perceptual model of hillslope flow at the Maimai catchment. Advances in Hillslope Processes, 1, 35-60.

Braun L N, Aellen M, Funk M, et al (1994). Measurement and simulation of high alpine water balance components in the Linth-Limmern head watershed (north-eastern Switzerland). Zeitschrift für Gletscherkunde und Glazialgeologie, 30, 161-185.

Bravo E L, Suárez M H, Cueto O G, et al (2012). Determination of basics mechanical properties in a tropical clay soil as a function of dry bulk density and moisture. Soil Water, 21, 5-11.

Bromhead E N. (1978). Large landslides in London Clay at Herne Bay, Kent. Quarterly Journal of Engineering Geology and Hydrogeology, 11(4), 291-304.

Burda J, Vilimek V. (2010). The influence of climate effects and fluctuations in groundwater level on the stability of anthropogenic foothill slopes in the Krusne Hory Mountains, Czechia. Geografie, 115(4), 377-392.

Carey J M. (2011). The progressive development and post-failure behaviour of deep-seated landslide complexes, Ph. D. Book, Durham University, United Kingdom

Carey J M. Petley D N (2014). Progressive shear-surface development in cohesive materials implications for landslide behaviour. Engineering Geology, 177, 54-65.

Carrubba P, Del Fabbro M. (2008). Laboratory investigation on reactivated residual strength. Journal of Geotechnical and Geoenvironmental Engineering, 134(3), 302.

Casadei M, Dietrich W E, Miller N L. (2003). Testing a model for predicting the timing and location of shallow landslide initiation in soil-mantled landscapes. Earth Surface Processes and Landforms, 28(9), 925-950.

Chow V T. (1964). Runoff, handbook of applied hydrology. Mc Graw Hill Book Company, New York, USA.

Chen L, Young M H. (2006). Green-Ampt infiltration model for sloping surfaces. Water Resources Research, 42(7), 1-9.

Claessens L, Knapen, A, Kitutu M G, et al (2007). Modelling landslide hazard, soil redistribution and sediment yield of landslides on the Ugandan footslopes of Mount Elgon. Geomorphology, 90(1), 23-35.

Collins B D, Znidarcic D. (2004). Stability analyses of rainfall induced landslides. Journal of Geotechnical and Geoenvironmental Engineering, 130(4), 362.

Collison A J C. Anderson M G. (1996). Using a combined slope hydrology/stability model to identify suitable conditions for landslide prevention by vegetation in the humid tropics. Earth Surface Processes and Landforms, 21(8), 737-747.

Corominas J, Moya J, Ledesma A, et al (2005). Prediction of ground displacements and velocities from groundwater level changes at the Vallcebre landslide (Eastern Pyrenees, Spain). Landslides, 2(2), 83-96.

Cruden D M. (1991). A simple definition of a landslide. Bulletin of the International Association of Engineering Geology, 43(1), 27-29.

Cruden D M, Varnes D J. (1996). Landslide types and processes. Turner AK, Schuster, RL Landslides Investigation and Mitigation, special report 247, (1982), 36-75.

Czys R R, Scott R W, Tang K C, et al. (1996). A physically based, nondimensional parameter for discriminating between locations of freezing rain and ice pellets. Weather and Forecasting, 11(4), 591-598.

Dafalla M A. (2013). Effects of clay and moisture content on direct shear tests for clay-sand mixtures. Advances in Materials Science and Engineering, vol. 2013, doi:10.1155/2013/562726.

Di Maio C, Vassallo R, Vallario M. (2013). Plastic and viscous shear displacements of a deep and very slow landslide in stiff clay formation. Engineering Geology, 162, 53-66.

Dingman S L. (1994). Physical hydrology. Macmillan Publishing Company, New York, USA.

Dikau R. (1996). Landslide recognition: identification, movement, and clauses, John Wiley & Sons, New Jersey, USA.

Doussan C, Ledoux L, Detay M. (1998). River-groundwater exchanges, bank filtration, and groundwater quality: Ammonium behavior. Journal of Environmental Quality, 27(6), 1418-1427.

Duncan J M, Wright S G, Brandon T L. (2014). Soil strength and slope stability, John Wiley & Sons, New Jersey, USA.

El-Maksoud M A F. (2006). Laboratory determining of soil strength parameters in calcareous soils and their effect on chiseling draft prediction. Proceedings of Energy Efficiency and Agricultural Engineering International Conference, Rousse.

Esko K. (1980). On the values and variability of degree-day melting factor in Finland. Nordic Hydrology, 11(5), 235-242.

Faris F, Fathani F. (2013). A coupled hydrology/slope kinematics model for developing early warning criteria in the kalitlaga landslide, Banjarnegara, Indonesia. Progress of Geo-Disaster Mitigation Technology in Asia. Springer Berlin Heidelberg, Germany.

Festl J, Singer J, Thuro K. (2011). The Aggenalm landslide-first findings of the acquired monitoring data. Proceedings of the International Symposium on Rock Slope Stability in Open Pit Mining and Civil Engineering. Vancouver.

Festl J. (2014). Analysis and evaluation of the geosensor network's data at the Aggenalm landslide, Bayerischzell, Germany. Ph. D. Book, Technical University Munich, Munich, Germany.

Finsterwalder S. (1887). Der suldenferner. Zeitschrift des Deutschen und Osterreichischen Alpenvereins, 18, 72-89.

Fredlund D G, Morgenstern N R, Widger R A. (1978). The shear strength of unsaturated soils. Canadian Geotechnical Journal, 15(3), 313-321.

Fredlund D G, Rahardjo H. (1993). Soil mechanics for unsaturated soils, John Wiley & Sons, New Jersey, USA.

Fredlund D G, Xing A. (1994). Equations for the soil-water characteristic curve. Canadian Geotechnical Journal, 31(6), 1026.

Freedman D, Pisani R, Purves, R. (1998). Statistics (3rd ed). Norton & Company, New York, USA.

Gao Y B, Ge X N, Song J. (2013). Strengths of Unsaturated Silty Clay Used as Garden Hill Fill in Shanghai. New Frontiers in Engineering Geology and the Environment. Springer Berlin Heidelberg,

Germany.

Garen D C, Marks D. (2005). Spatially distributed energy balance snowmelt modelling in a mountainous river basin: Estimation of meteorological inputs and verification of model results. Journal of Hydrology, 315, 126-153.

Gallemann T. (2012). Geodätische Überwachung der Rutschung Aggenalm im Sudelfeld, 9. Folgemessung-LfU unpublished internal report, 9.

Giannecchini R, Galanti Y, D'Amato Avanzi G. (2012). Critical rainfall thresholds for triggering shallow landslides in the Serchio River Valley (Tuscany, Italy). Natural Hazards and Earth System Science, 12, 829-842.

Gibo S, Egashira K, Ohtsubo M, et al(2002). Strength recovery from residual state in reactivated landslides. Géotechnique, 52(9), 683-686.

Gonzalez D A, Ledesma A, Corominas J. (2008). The viscous component in slow moving landslides: a practical case. Landslides and Engineering Slopes, 237-242.

Gollnitz W D. (2003). Infiltration rate variability and research needs. Riverbank Filtration. Springer Berlin Heidelberg, Germany.

Gottardi G, Butterfield R. (2001). Modelling ten years of downhill creep data. Proceedings of the 15th International Conference on Soil Mechanics and Geotechnical Engineering, Istanbul, Turkey.

Green W H, Ampt G A. (1911). Studies on soil physics, 1. The flow of air and water through soils. Journal of Agricultural Science, 4(1), 1-24.

Guzzetti F, Peruccacci, S, Rossi M, et al(2007). Rainfall thresholds for the initiation of landslides in central and southern Europe. Meteorology and Atmospheric Physics, 98(3-4), 239-267.

Guzzetti F, Peruccacci S, Rossi M, et al(2008). The rainfall intensity-duration control of shallow landslides and debris flows: an update. Landslides, 5(1), 3-17.

Gwinner M P. (1971). Geologie der Alpen. - 477 pp. Stuttgart (Schweizerbart).

Häggmark L, Ivarsson K I, Olofsson P O. (1997). MESAN: mesoskalig analys.

Herrmann A. (1978). Schneehydrologische Untersuchungen in einem randalpinen Niederschlagsgebiet (Lainbachtal bei Benediktbeuren/Oberbayern. -Münchner Geograph. Abh., Bd. 22, Inst. f. Geographie, München.)

Herrera G, Fernández-Merodo J A, Mulas J, et al(2009). A landslide forecasting model using ground based SAR data: The Portalet case study. Engineering Geology, 105(3), 220-230.

Herrero J, Polo M J, Moñino A, et al(2009). An energy balance snowmelt model in a Mediterranean site. Journal of Hydrology, 371(1), 98-107.

Hock R. (1999). A distributed temperature-index ice- and snowmelt model including potential direct solar radiation. Journal of Glaciology, 45(149), 101-111.

Hock R. (2003). Temperature index melt modelling in mountain areas. Journal of Hydrology, 282 (1), 104-115.

Hong Y, Hiura H, Shino K, et al(2005). The influence of intense rainfall on the activity of large-scale crystalline schist landslides in Shikoku Island, Japan. Landslides, 2(2), 97-105.

Houlsby G T. (1991). How the dilatancy of soils affects their behaviour, University of Oxford, Department of Engineering Science, UK.

Hornberger G M, Germann P F, Beven K J. (1991). Through flow and solute transport in an isolated sloping soil block in a forested catchment. Journal of Hydrology, 124(1), 81-99.

Huang A B, Lee J T, Ho Y T, et al(2012). Stability monitoring of rainfall-induced deep landslides through pore pressure profile measurements. Soils and Foundations, 52(4), 737-747.

Hungr O, Evans S G, Bovis M J, et al(2001). A review of the classification of landslides of the flow type. Environmental & Engineering Geoscience, 7(3), 221-238.

Hutchinson J N. (1988). General Report: Morphological and geotechnical parameters of landslides in relation to geology and hydrogeology. Process of 5th International Symposium on Landslides, Lausanne.

Ishihara Y, Kobatake S. (1979). Runoff model for flood forecasting. Bulletin of the Disaster Prevention Research Institute, 29 (1), 27-43.

Iverson R M. (1997). The physics of debris flows. Reviews of Geophysics, 35(3), 245-296.

Iverson R M. (2000). Landslide triggering by rain infiltration. Water Resources Research, 36(7), 1897-1910.

Iverson R M. (2005). Regulation of landslide motion by dilatancy and pore pressure feedback. Journal of Geophysical Research: Earth Surface, 110(2), 1-16.

Johnson K A, Sitar N. (1990). Hydrologic conditions leading to debris-flow initiation. Canadian Geotechnical Journal, 27(6), 789-801.

Jost G, Moore R D, Smith R, et al(2012). Distributed temperature-index snowmelt modelling for forested catchments. Journal of Hydrology, 420, 87-101.

Katte V Y, Blight G E. (2015). Solute suction and shear strength in saturated soils. Advances in Unsaturated Soils, Taylor & Francis Group, London, UK.

Keefer D K, Johnson A M. (1983). Earth flows; morphology, mobilization, and movement, United States Government Printing Office, Washington, USA.

Kilsby C. (2007). An engineering geological appraisal of the Utiku Landslide, North Island, New Zealand. M. Sc. Book, University of Portsmouth, Portsmouth, United Kingdom.

Kimura S, Nakamura S, Vithana S B, et al(2014). Shearing rate effect on residual strength of landslide soils in the slow rate range. Landslides, 11(6), 969-979.

Kustas W P, Rango A, Uijlenhoet R. (1994). A simple energy budget algorithm for the snowmelt runoff model. Water Resources Research, 30(5), 1515-1527.

Lakhankar T Y, Muñoz J, Powell A M, et al (2013). CREST-Snow Field Experiment: analysis of snowpack properties using multi-frequency microwave remote sensing data. Hydrology and Earth System Sciences, 17(2), 783-793.

Lateltin O, Beer C, Raetzo H, et al(1997). Landslides in flysch terranes of Switzerland: causal factors and climate change. Eclogae Geologicae Helvetiae, 90(3), 401-406.

Lemos L, Skempton A W, Vaughan P R. (1985). Earthquake loading of shear surfaces in slope, Proceedings of the 11th International Conference on Soil Mechanics and Foundation Engineering, San Francisco.

Lee J M, Bland K J, Townsend D B, et al(2012). Geology of the Hawke's Bay area. Institute of Geological and Nuclear Sciences 1: 250000 geological map 8. Lower Hutt. Institute of Geological and Nuclear Sciences Limited.

Leroueil S, Locat J, Vaunat J, et al(1996). Geotechnical characterization of slope movements. Proceedings of the 7th International Symposium on Landslides, Trondheim.

Li D, Yin K, Leo C. (2010). Analysis of Baishuihe landslide influenced by the effects of reservoir water and rainfall. Environmental Earth Sciences, 60(4), 677-687.

Lora M, Camporese M, Salandin P. (2016). Design and performance of a nozzle-type rainfall simulator for landslide triggering experiments. Catena, 140, 77-89.

Lucas A, Mangeney A, Ampuero J P. (2014). Frictional velocity-weakening in landslides on earth and on other planetary bodies. Nature Communications, 5, 3417.

Lupini J F, Skinner A E, Vaughn P R. (1981). The drained residual strength of cohesive soils: Géotechnique, 31(2), 181-213.

Madritsch H, Millen B M J. (2007). Hydrogeologic evidence for a continuous basal shear zone within a deep-seated gravitational slope deformation (Eastern Alps, Tyrol, Austria). Landslides, 4(2), 149-162.

Marklseder F. (1935). Bergrutsch am Tatzelwurm. -Anzeiger für Oberaudorf und Kiefersfelden-Heimatzeitung des Inngaues, 26(17), 2.

Malaise von V. (1951). Die Landwirtschaft des Landkreises Rosenheim im allgemeinen, die Almwirtschaft im besonderen und die Massnahmen zu ihrer Förderung. Ph. D. Book, Technical University Munich, Munich, Germany.

Malet J P, Van Asch T W, Van Beek R, et al(2005). Forecasting the behaviour of complex landslides with a spatially distributed hydrological model. Natural Hazards and Earth System Science, 5(1), 71-85.

Manzari M T, Nour M A. (2000). Significance of soil dilatancy in slope stability analysis. Journal of Geotechnical and Geoenvironmental Engineering, 126(1), 75-80.

Massey C I. (2010). The dynamics of reactivated landslides: Utiku and Taihape, North Island, New Zealand. PhD Book, Durham University, United Kingdom.

Massey C I., Petley D N, McSaveney M J. (2013). Patterns of movement in reactivated landslides. Engineering Geology, 159, 1-19.

Matsuo T, Sasyo Y. (1981). Non-melting phenomena of snowflakes observed in subsaturated air below freezing level. Journal of the Meteorological Society of Japan, 59, 26-32.

Matsushi Y, Matsukura Y. (2006). Cohesion of unsaturated residual soils as a function of volumetric water content. Bulletin of Engineering Geology and the Environment, 65(4), 449-455.

Matsuura S, Asano S, Okamoto T, et al(2003). Characteristics of the displacement of a landslide with shallow sliding surface in a heavy snow district of Japan. Engineering Geology, 69(1), 15-35.

Matsuura S, Asano S, Okamoto, T. (2008). Relationship between rain and/or meltwater, pore-water pressure and displacement of a reactivated landslide. Engineering Geology, 101(1-2), 49-59.

Mayer K, Müller-Koch K, von Poschinger, A. (2002). Dealing with landslide hazards in the Bavarian Alps. Proceedings of the 1st European conference on landslides, Prague.

McDonnell J J., Sivapalan M, Vaché K, et al(2007). Moving beyond heterogeneity and process complexity: A new vision for watershed hydrology. Water Resources Research, 43(7), W07301.

McKyes E, Nyamugafata P, Nyamapfene K W. (1994). Characterization of cohesion, friction and sensitivity of two hardsetting soils from Zimbabwe. Soil and Tillage Research, 29(4), 357-366.

Mesri G, Huvaj-Sarihan N. (2012). Residual shear strength measured by laboratory tests and mobilized in landslides. Journal of Geotechnical and Geoenvironmental Engineering, 138(5), 585-593.

Miao H, Wang G, Yin K, et al (2014). Mechanism of the slow-moving landslides in Jurassic red-strata in the Three Gorges Reservoir, China. Engineering Geology, 171, 59-69.

Michiue M. (1985). A method for predicting slope failures on cliff and mountain due to heavy rain. Natural Disaster Science, 7(1), 1-12.

Moore R, Lee E M, Clark A R. (1995). The Undercliff of the Isle of Wight: a review of ground behaviour. Report by Rendel Geotechnics for South Wight Borough Council, Ventnor, Isle of Wight.

Moore P L, Iverson N R. (2002). Slow episodic shear of granular materials regulated by dilatant strengthening. Geology, 30(9), 843-846.

Moore R, Carey J, Mills A, et al(2006). Recent landslide impacts on the UK Scottish road network: investigation into the mechanisms, causes and management of landslide risk. Proceedings of the International Conference on Slopes, Bangkok.

Moore R, Carey J M, McInnes R G, et al(2007a). Climate change, so what? Implications for ground movement and landslide event frequency in the Ventnor Undercliff, Isle of Wight. Landslides and Climate Change: Challenges and Solutions. Balkema, Rotterdam, Netherland.

Moore R, Turner M D, Palmer M J, et al(2007b). The Ventnor Undercliff: Landslide model, mechanisms and causes, and the implications of climate change induced ground behaviour and risk. Proceedings of the International Conference on Landslides and Climate Change. London.

Moore R Carey J M, McInnes R G. (2010). Landslide behaviour and climate change: predictable consequences for the Ventnor Undercliff, Isle of Wight. Quarterly Journal of Engineering Geology and

Hydrogeology, 43(4), 447-460.

Moriasi D N, Arnold J G, Van Liew M W, et al (2007). Model evaluation guidelines for systematic quantification of accuracy in watershed simulations. Transactions of the ASABE, 50(3), 885-900.

Mutiti S, Levy J. (2010). Using temperature modeling to investigate the temporal variability of riverbed hydraulic conductivity during storm events. Journal of Hydrology, 388(3-4), 321-334.

Murphy A. (1988). Skill scores based on the mean square error and their relationships to the correlation coefficient. Monthly Weather Review 116, 2417-2424.

Nash J E, Sutcliffe J V. (1970). River flow forecasting through. Part I. A conceptual models discussion of principles. Journal of Hydrology. 10, 282-290.

Nakai T, Hinokio M. (2004). A simple elastoplastic model for normally and over consolidated soils with unified material parameters. Soils and Foundations, 44(2), 53-70.

Nickmann M, Spaun G, Thuro K. (2006). Engineering geological classification of weak rocks. International Association for Engineering Geology and the Environment, 492, 9.

Nishii R, Matsuoka N. (2010). Monitoring rapid head scarp movement in an alpine rockslide. Engineering Geology, 115(1-2), 49-57.

Noverraz F, Bonnard C. (1998). Grands glissements de versants et climat, vdf.

Ochepo J, Stephen O D, Masbeye O. (2012). Effect of water cement ratio on cohesion and friction angle of expansive black clay of Gombe State, Nigeria. Electronic Journal of Geotechnical Engineering, 17, 931-945.

Ohtsu H, Janrungautai S, Takahashi K. (2003). A study on the slope risk evaluation due to rainfall using the simplified storage tank model. Proceeding of the 2nd Southeast Asia Workshop on Rock Engineering, Bangkok, Thailand,.

Pack R T. (2001). Assessing Terrain Stability in a GIS using SINMAP. 15th Annual GIS Conference, Vancouver.

Petley D N, Higuchi T, Petley D J, et al (2005). Development of progressive landslide failure in cohesive materials. Geology, 33(3), 201-204.

Picarelli L, Urciuoli G, Russo C. (2004). Effect of groundwater regime on the behaviour of clayey slopes. Canadian Geotechnical Journal, 41, 467-484.

Picarelli L. (2007). Considerations about the mechanics of slow active landslides in clay. Progress in Landslide Science, Springer Berlin Heidelberg, Germany.

Pillans B. (1986). A Late Quaternary uplift map for North Island, New Zealand. Royal Society of New Zealand, Bulletin 24, 409-417.

Pinault J L, Schomburgk S. (2006). Inverse modeling for characterizing surface water/groundwater exchanges. Water Resources Research, 42(8), W08414.

Pulford A, Stern T. (2004). Pliocene exhumation and landscape evolution of central North Island, New Zealand: the role of the upper mantle. Journal of Geophysical Research: Earth Surface, 109

(F1).

Rahardjo H, Lim T T, Chang M F, et al (1995). Shear-strength characteristics of a residual soil. Canadian Geotechnical Journal, 32(1), 60-77.

Rahardjo H, Leong E C, Rezaur R B. (2008). Effect of antecedent rainfall on pore-water pressure distribution characteristics in residual soil slopes under tropical rainfall. Hydrological Processes, 22, 506-523.

Rahardjo H, Nio A S, Leong E C, et al (2010). Effects of groundwater table position and soil properties on stability of slope during rainfall. Journal of Geotechnical and Geoenvironmental Engineering, 136(11), 1555-1564.

Ramer J. (1993). An empirical technique for diagnosing precipitation type from model output. 5th International Conference on Aviation Weather Systems, Vienna.

Ranalli M, Gottardi G, Medina-Cetina Z, et al (2010). Uncertainty quantification in the calibration of a dynamic viscoplastic model of slow slope movements. Landslides, 7(1), 31-41.

Rango A, Martinec J. (1995). Revisiting the degree-day method for snowmelt computations. Water Resources Bulletin, 31(4), 657-669.

Richards L A. (1931). Capillary conduction of liquids through porous mediums. Journal of Applied Physics, 1, 318-333.

Rogers N W, Selby M J. (1980). Mechanisms of shallow translational landsliding during summer rainstorms: North Island, New Zealand. Geografiska Annaler. Series A. Physical Geography, 11-21.

Rosenberry D O, Healy R W. (2012). Influence of a thin veneer of low-hydraulic-conductivity sediment on modelled exchange between river water and groundwater in response to induced infiltration. Hydrological Processes, 26(4), 544-557.

Schaap M G, van Genuchten M T. (2006). A Modified Mualem-van Genuchten Formulation for Improved Description of the Hydraulic Conductivity Near Saturation. Vadose Zone Journal, 5 (1), 27.

Schaeffer D G, Iverson R M. (2008). Steady and intermittent slipping in a model of landslide motion regulated by pore-pressure feedback. SIAM Journal on Applied Mathematics, 69(3), 769-786.

Schulz W H, McKenna J P, Kibler J D, et al (2009). Relations between hydrology and velocity of a continuously moving landslide—evidence of pore-pressure feedback regulating landslide motion? Landslides, 6(3), 181-190.

Schulz W H, Wang G. (2014). Residual shear strength variability as a primary control on movement of landslides reactivated by earthquake-induced ground motion: Implications for coastal Oregon, US. Journal of Geophysical Research: Earth Surface, 119(7), 1617-1635.

Schmidt-thome P. (1964). Alpenraum. - In: BAYERISCHES GEOLOGISCHES LANDESAMT (LFU) [ed.]. Erläuterungen geol. Kt. Bayern 1:500000, 2nd ed. - 344 pp., Munich (LfU), 244-297.

Secondi M M, Crosta G, di Prisco C, et al (2013). Landslide motion forecasting by a dynamic viscoplastic model. Landslide science and practice. Springer Berlin Heidelberg.

Shimizu M. (1982). Effect of overconsolidation on dilatancy of a cohesive soil. Soils & Foundations, 22(4), 121-135.

Sidle R C. (2006). Field observations and process understanding in hydrology: Essential components in scaling. Hydrological Processes, 20(6), 1439-1445.

Simoni A, Berti M, Generali M, et al(2004). Preliminary result from pore pressure monitoring on an unstable clay slope. Engineering Geology, 73(1), 117-128.

Singer J, Schuhbäck S, Wasmeier P et al. (2009). Monitoring the aggenalm landslide using economic deformation measurement Techniques. Austrian Journal of Earth Sciences, 102(2), 20-34.

Sivapalan M, Jothityangkoon C, Menabde M. (2002). Linearity and nonlinearity of basin response as a function of scale: discussion of alternative definitions. Water Resources Research, 38(2).

Skempton A W. (1970). First-time slides in over-consolidated clays. Géotechnique, 20(3), 320-324.

Skempton A W. (1985). Residual strength of clays in landslides, folded strata and the laboratory. Géotechnique, 35(1), 3-18.

Stark T D, Hussain M. (2010). Drained residual strength for landslides. GeoFlorida, 3217-3226.

Suzuki M, Kobashi S. (1981). The critical rainfall for the disasters caused by slope failures. Journal of Japan Society of Erosion Control Engineering (Shin-Sabo), 34(2), 16-26.

Swanston D N, Ziemer R R, Janda R J. (1995). Rate and mechanics of progressive hillslope failure in the Redwood Creek basin, northwestern California, Miscellaneous Publication, Washington, USA.

Takahashi K. (2004). Research of underground water numerical analysis method that considering water circulation system, Ph. D. Book, Kyoto university, Japan.

Takahashi K, Ohnish Y, Xiong, J, et al. (2008). Tank model and its application to groundwater table prediction of slope. Chinese Journal of Rock Mechanics and Engineering, 27(12), 2501-2508.

Talebi A, Uijlenhoet R, Troch P A. (2007). Soil moisture storage and hillslope stability. Natural Hazards and Earth System Science, 7, 523-534.

Thiebes B, Bell R, Glade T, et al(2014). Integration of a limit-equilibrium model into a landslide early warning system. Landslides, 11(5), 859-875.

Thuro K, Wunderlich Th. Heunecke O, et al(2009). Low cost 3D early warning system for alpine instable slopes-the Aggenalm Landslide monitoring system. Geomechanics & Tunnelling, 3, 221-237.

Thuro K, Singer J, Festl J, et al (2010). New landslide monitoring techniques-developments and experiences of the alpEWAS project. Journal of Applied Geodesy, 4, 69-90.

Thuro K, Singer J, Festl J. (2011a). A geosensor network based monitoring and early warning system for landslides. The Second World Landslide Forum, Roma, Italy.

Thuro K, Singer J, Festl J. (2011b). Low cost 3D early warning system for alpine instable slopes-the Aggenalm Landslide monitoring system. International Symposium on Rock Slope Stability in Open Pit Mining and Civil Engineering, Vancouver.

Thuro K, Singer J, Festl J. (2013). A Geosensor Network Based Monitoring and Early Warning System for Landslides. Landslide Science and Practice, Volume 2: Early Warning, Instrumentation and Monitoring. Heidelberg, New York.

Tika TH E, Hutchinson J N. (1999). Ring shear tests on soil from the Vaiont landslide slip surface. Géotechnique, 49(1), 59-74.

Trémolières M, Eglin I, Roeck U, et al (1993). The exchange process between river and groundwater on the Central Alsace floodplain (Eastern France). Hydrobiologia, 254(3), 133-148.

Uchimura T, Towhata I, Anh T T L, et al (2010). Simple monitoring method for precaution of landslides watching tilting and water contents on slopes surface. Landslides, 7(3), 351-357.

Waller G n. d. : Die Brücke-Erlebtes und Erlauschtes vom Leben auf dem Berg. - 19 pp.

Wagner A J. (1957). Mean temperature from 1000 mb to 500 mb as a predictor of precipitation type, MIT Department of Meteorology, USA.

Wakizaka Y. (2013). Characteristics of crushed rocks observed in drilled cores in landslide bodies located in accretionary complexes. Tectonophysics, 605, 114-132.

Wang G, Sassa K. (2003). Pore-pressure generation and movement of rainfall-induced landslides: effects of grain size and fine-particle content. Engineering Geology, 69, 109-125.

Wang G, Suemine A, Schulz W H. (2010). Shear-rate-dependent strength control on the dynamics of rainfall-triggered landslides, Tokushima Prefecture, Japan. Earth Surface Processes and Landforms, 35(4), 407-416.

Vanapalli S K, Fredlund D G, Pufahl D E, et al (1996). Model for the prediction of shear strength with respect to soil suction. Canadian Geotechnical Journal, 33(3), 379-392.

Van Asch T W J, Buma J, Van Beek L P H. (1999). A view on some hydrological triggering systems in landslides. Geomorphology, 30(1-2), 25-32.

Van Asch T W J, Van Beek L P H, Bogaard T A. (2007). Problems in predicting the mobility of slow-moving landslides. Engineering Geology, 91(1), 46-55.

Van Asch T W J, Malet J P, Bogaard T A. (2009). The effect of groundwater fluctuations on the velocity pattern of slow-moving landslides. Natural Hazards and Earth System Science, 9(3), 739-749.

Van Beek L P H, Van Asch T W. (2004). Regional assessment of the effects of land-use change on landslide hazard by means of physically based modelling. Natural Hazards, 31(1), 289-304.

Van Genuchten, (1980). A closed-form equation for predicting the hydraulic conductivity of unsaturated soils. Soil Science Society of America Journal, 44(5), 892-898.

Weiler M, McDonnell J J, Tromp-van Meerveld I, et al (2005). Subsurface stormflow. Encyclopedia of

Hydrological Sciences, 3, 1-14.

Weill S, Mouche E, Patin J. (2009). A generalized Richards equation for surface/subsurface flow modelling. Journal of Hydrology, 366(1), 9-20.

Wienhofer J, Germer K, Lindenmaier F, et al (2009). Applied tracers for the observation of subsurface stormflow at the hillslope scale. Hydrology and Earth System Sciences, 13(7), 1145.

Wilkinson P L, Anderson M G, Lloyd D M. (2002). An integrated hydrological model for rain-induced landslide prediction. Earth Surface Processes and Landforms, 27(12), 1285-1297.

Vlcko J, Greif V, Grof V, et al (2009). Rock displacement and thermal expansion study at historic heritage sites in Slovakia. Environmental Geology, 58(8), 1727-1740.

Wolff H. (1985). Geologische Karte von Bayern 1 : 25000, Erläuterungen zum Blatt Nr. 8338 Bayrischzell. - 190 pp., Munich (Bayer. LfU).

Wwa rosenheim-wasserwirtschaftsamt rosenheim n. d. : Talzuschub Gassenbach - Aggraben, Östliches Sudelfeld Gemeinde Oberaufdorf. - internal report, 2 pp.

Vulliet L, Hutter K. (1988). Viscous-type sliding laws for landslides. Canadian Geotechnical Journal, 25(3), 467-477.

Xiong J, Ohnish Y, Takahashi K, et al (2009). Parameter determination of multi-tank model with dynamically dimensioned search. Process Symposium Rock Mechanics Japan, Kyoto.

Xu L M, Zhu H H, Nakai T, et al (2006). Numerical simulation of shear band in overconsolidated clay. Rock and Soil Mechanics, 27(1), 61.

Yin Y, Wang H, Gao Y, et al (2010). Real-time monitoring and early warning of landslides at relocated Wushan Town, the Three Gorges Reservoir, China. Landslides, 7(3), 339-349.

Appendix I   Codes of landslide model

Appendix I(UK case) displays (1) Cohesion back analysis in quasi-static landslide model. (2) Quasi-static landslide prediction (cohesion constant). (3) Quasi-static landslide prediction (cohesion-velocity simple linear).

1 Cohesion back analysis in quasi-static landslide model

```
clear
load('matlab1.mat');
ut = pwp;
n = 0.2;
r1 = 1900;
tanb = 0.214;
c = 45;
v = 4.7E+11;
m = 20;

hb = cell(1,5);
hb1 = [65,120];
hb2 = [120,115,110,105,85,80,75, 70];
hb3 = [65, 55,55,45,40, 35, 35, 70];
hb4 = [70,70, 70, 75, 80, 70, 45,40];
hb5 = [40,15,15,15,15,15,15];

hb{1} = hb1;
hb{2} = hb2;
hb{3} = hb3;
hb{4} = hb4;
```

hb{5} = hb5;

cosa1 = [0.089, 0.089];
cosa2 = [0.992, 0.992, 0.992, 0.996, 0.992, 0.992, 0.992, 0.992];
cosa3 = [0.992, 0.992, 0.992, 0.992, 0.992, 0.992, 0.992, 0.992];
cosa4 = [0.992, 0.992, 0.992, 0.992, 0.992, 0.992, 0.992, 0.992];
cosa5 = [0.992, 0.992, 0.992, 0.992, 0.992, 0.992, 0.992];
cosa{1} = cosa1;
cosa{2} = cosa2;
cosa{3} = cosa3;
cosa{4} = cosa4;
cosa{5} = cosa5;

sina1 = [0.37, 0.37];
sina2 = [0.122, 0.122, 0.122, 0.122, 0.122, 0.122, 0.122, 0.122];
sina3 = [0.122, 0.122, 0.122, 0.122, 0.122, 0.122, 0.122, 0.122];
sina4 = [0.122, 0.122, 0.122, 0.122, 0.122, 0.122, 0.122, 0.122];
sina5 = [0.122, 0.122, 0.122, 0.122, 0.122, 0.122, 0.122];
sina{1} = sina1;
sina{2} = sina2;
sina{3} = sina3;
sina{4} = sina4;
sina{5} = sina5;

% giving initial parameter(ut is pore water pressure; n is the porosity of slope mass; r1 is the density of slope mass; b is the friction angle of materials; c is the cohesion; v is the viscous coefficient; m is the width of slice; hbi is the slice hight; ai is the slope angle ($i = 1 \cdots 5$))

N = length(hb);
y = zeros(size(ut));
y_sum = zeros(size(y));

```
for t=2:length(ut);
y(t)= displacement(t+1)-displacement(t);
r=(80*r1+(ut(t))*100*n)/80;
% density dependent on availability of water
    y(1)= 1.16E-09;
    a(1)= 2.53e-09;
    for ii=1:N
        fb1(ii)= sum (10*r*m*hb{ii}.*cosa{ii}-20000*(510+ut(t)))
        *tanb;
        fb2(ii)= sum(10*r*m*hb{ii}.*sina{ii});
end
    fbr(t)= sum(fb1(:));
    fbs(t)= sum(fb2(:));
a(t)= y(t)-y(t-1);
fbc(t) = fbs(t)-fbr(t)-75810000*a(t-1)-v*y(t);
    if fbc(t)<0
        fbc(t)= 0;
    end
    y_sum(t)= y_sum(t-1)+y(t);
    a(t)= y(t)-y(t-1);
end
plot(fbc,'-r')
```

% cohesion calculation(where fbc is the cohesion; $a(t)$ is the acceleration at time t; $y(t)$ is the displacement at time t; y_sum(t) is the cumulative displacement at time t)

2 Quasi-static landslide prediction (cohesion constant)

```
clear all;
clc;

load('matlab');
num=1400;
```

```
ut = pwp;
n = 0.2;
r1 = 1900;
tanb = 0.217;
v = 4.7E+11;

m = 20;

hb = cell(1, 5);
hb1 = [65, 120];
hb2 = [120, 115, 110, 105, 85, 80, 75, 70];
hb3 = [65, 55, 55, 45, 40, 35, 35, 70];
hb4 = [70, 70, 70, 75, 80, 70, 45, 40];
hb5 = [40, 15, 15, 15, 15, 15, 15];

hb{1} = hb1;
hb{2} = hb2;
hb{3} = hb3;
hb{4} = hb4;
hb{5} = hb5;

cosa1 = [0.089, 0.089];
cosa2 = [0.992, 0.992, 0.992, 0.996, 0.992, 0.992, 0.992, 0.992];
cosa3 = [0.992, 0.992, 0.992, 0.992, 0.992, 0.992, 0.992, 0.992];
cosa4 = [0.992, 0.992, 0.992, 0.992, 0.992, 0.992, 0.992, 0.992];
cosa5 = [0.992, 0.992, 0.992, 0.992, 0.992, 0.992, 0.992];
cosa{1} = cosa1;
cosa{2} = cosa2;
cosa{3} = cosa3;
cosa{4} = cosa4;
cosa{5} = cosa5;
```

```
sina1 = [0.37, 0.37];
sina2 = [0.122, 0.122, 0.122, 0.122, 0.122, 0.122, 0.122, 0.122];
sina3 = [0.122, 0.122, 0.122, 0.122, 0.122, 0.122, 0.122, 0.122];
sina4 = [0.122, 0.122, 0.122, 0.122, 0.122, 0.122, 0.122, 0.122];
sina5 = [0.122, 0.122, 0.122, 0.122, 0.122, 0.122, 0.122, 0.122];
sina{1} = sina1;
sina{2} = sina2;
sina{3} = sina3;
sina{4} = sina4;
sina{5} = sina5;
% giving initial parameter(ut is pore water pressure; n is the porosity of slope mass;
% r1 is the density of slope mass; b is the friction angle of materials; c is the cohesion; v
% is the viscous coefficient; m is the width of slice; hbi is the slice hight; ai is the slope
% angle (i=1...5))

N = length(hb);
y = zeros(size(ut));

y_sum = zeros(size(y));

for t = 2:length(ut);
    r = (80 * r1+(ut(t))) * 100 * n)/80;
% density dependent on availability of water

y(1) = 1.16E-09;
a(1) = 2.53e-09;
for ii = 1:N

fb1(ii) = sum (10 * r * m * hb{ii}. * cosa{ii}-20000 * (510+ut(t))) * tanb;
fb2(ii) = sum(10 * r * m * hb{ii}. * sina{ii});
end
fbr(t) = sum(fb1(:));
```

fbs(t) = sum(fb2(:));

y(t) = ((fbs(t)-fbr(t)-(3.5834E+07)-75810000*a(t-1))/v);
if y(t)<0
y(t) = 0;
end
y_sum(t) = y_sum(t-1)+y(t);
a(t) = y(t)-y(t-1);
end
% displacement prediction (where a(t) is the acceleration at time t; y(t) is displacement (y(t) = velocity, when considering unit is day) at time t; y_sum(t) is the cumulative displacement at time t;)
plot(y_sum,'-r')
hold on
plot(displacement);

3 Quasi-static landslide prediction (cohesion-velocity simple linear)

clear all;
clc;

load('matlab');
num = 1400;
ut = pwp;
n = 0.2;
r1 = 1900;
tanb = 0.217;
v = 4.7E+11;

m = 20;

hb = cell(1, 5);
hb1 = [65, 120];

```
hb2 = [120, 115, 110, 105, 85, 80, 75, 70];
hb3 = [65, 55, 55, 45, 40, 35, 35, 70];
hb4 = [70, 70, 70, 75, 80, 70, 45, 40];
hb5 = [40, 15, 15, 15, 15, 15, 15];

hb{1} = hb1;
hb{2} = hb2;
hb{3} = hb3;
hb{4} = hb4;
hb{5} = hb5;

cosa1 = [0.089, 0.089];
cosa2 = [0.992, 0.992, 0.992, 0.996, 0.992, 0.992, 0.992, 0.992];
cosa3 = [0.992, 0.992, 0.992, 0.992, 0.992, 0.992, 0.992, 0.992];
cosa4 = [0.992, 0.992, 0.992, 0.992, 0.992, 0.992, 0.992, 0.992];
cosa5 = [0.992, 0.992, 0.992, 0.992, 0.992, 0.992, 0.992];
cosa{1} = cosa1;
cosa{2} = cosa2;
cosa{3} = cosa3;
cosa{4} = cosa4;
cosa{5} = cosa5;

sina1 = [0.37, 0.37];
sina2 = [0.122, 0.122, 0.122, 0.122, 0.122, 0.122, 0.122, 0.122];
sina3 = [0.122, 0.122, 0.122, 0.122, 0.122, 0.122, 0.122, 0.122];
sina4 = [0.122, 0.122, 0.122, 0.122, 0.122, 0.122, 0.122, 0.122];
sina5 = [0.122, 0.122, 0.122, 0.122, 0.122, 0.122, 0.122];
sina{1} = sina1;
sina{2} = sina2;
sina{3} = sina3;
sina{4} = sina4;
sina{5} = sina5;
```

3 Quasi-static landslide prediction (cohesion-velocity simple linear).

% giving initial parameter (ut is pore water pressure; n is the porosity of slope mass; r1 is the density of slope mass; b is the friction angle of materials; c is the cohesion; v is the viscous coefficient; m is the width of slice; hbi is the slice hight; ai is the slope angle ($i=1\cdots5$)))

```
N = length(hb);
y = zeros(size(ut));
y_sum= zeros(size(y));

for t=2:length(ut);
    r=(80*r1+(ut(t))*100*n)/80;
% density dependent on availability of water

    y(1)= 1.16E-09;
    a(1)= 2.53e-09;
    for ii=1:N

        fb1(ii)= sum (10*r*m*hb{ii}.*cosa{ii}-20000*(510+ut(t)))*tanb;
        fb2(ii)= sum(10*r*m*hb{ii}.*sina{ii});
    end
        fbr(t)= sum(fb1(:));
        fbs(t)= sum(fb2(:));

y(t)= ((fbs(t)-fbr(t)-(-(3.354E+09)y1(t-1)+3.5834E+07)-75810000*a(t-1))/v);
    if y(t)<0
        y(t)= 0;
    end
    y_sum(t)= y_sum(t-1)+y(t);
    a(t)= y(t)-y(t-1);
end
```

% displacement prediction (where $a(t)$ is the acceleration at time t; $y(t)$ is displacement ($y(t)$ = velocity, when considering unit is day) at time t; $y\_sum(t)$ is the cumulative displacement at time t; $y1(t\text{-}1)$ is the real velocity of previous day)

plot(y_sum,'-r')

hold on

　plot(displacement);

Appendix II List of symbols

This list defines symbols used in the text.

Chapter 2

η	viscosity, Pa·s
ϕ'	effective friction angle, (°)
γ	unit weight of soil mass, kN/m³
a_c	acceleration, m/s²
α	slip surface inclination, (°)
a	relation parameter, 1
b	relation parameter, 1
c	relation parameter, 1
c'	effective soil cohesion, kPa
d	relation parameter, 1
F	slip force, kN
H	depth of slope surface to the potential failure surface, m
hs_i	upper surface water of ith day, mm
h'_{i+1}	lower ground water table of $i+1$th day, mm
h'_i	lower ground water table of ith day, mm
h_{i+1}	ground water table of $i+1$th day, mm
h_i	ground water table of ith day, mm
i_i	upper infiltration of ith day, mm

k	rate constant, 1
m	quality of the shear zone, kg
p_w	pore water pressure at the potential slip surface, kPa
Q	outflow, mm
q_i	drainage of ith day, mm
qs_i	upper surface runoff of ith day, mm
r_i	rainfall of ith day, mm
S	storage amount of water, mm
t	time, 1
T	indicates the buffering capacity of a reservoir or the "slowness" of water release, 1
T'	resistance force, kN
v	velocity, m/s
Z	height of the shear zone, m

Chapter 3

α	related coefficient between equivalent infiltration and increased ground water table, 1
α'	related coefficient between equivalent infiltration and increased pore water pressure, kPa/mm
β	average value of pore water pressure changed by drainage and ground water supply, kPa
a	related coefficient between pore water pressure of ith day and $i+1$th day without infiltration, kPa
b	related coefficient between pore water pressure of ith day and $i+1$th day without infiltration, 1
ER_i	equivalent rainfall of ith day, mm

ES_i	equivalent snowmelt of ith day, mm
f_m	degree-day factor for snowmelt rate, mm/°C
F	percentage of canopy cover, 1
g_i	ground water supply of ith day, mm
g'	acceleration of gravity, m/s²
h_i	ground water table height the ith day, mm
H	base water table, mm
M'	daily snowmelt, mm
n	average porosity of slope mass, 1
q_i	drainage of ith day, mm
PWP_i	pore water pressure of ith day, kPa
$\Delta PWP_{(g+q)i}$	PWP changed by drainage combined groundwater supply, kPa
ΔPWP_i	change of pore water pressure of ith day, kPa
$R_i^{(n)}$	part of rainfall of ith day changed to pore water pressure of ith day, mm
RH	relative humidity, 1
RH_t	threshold of relative humidity, 1
M	time until the effect of infiltration is reduced to 50%, 1
R_i	rainfall of ith day, mm
$S_i^{(n)}$	part of snowmelt of ith day to changed pore water pressure of ith day, mm
S_i	rainfall of ith day, mm
T_d	daily average temperature, °C

Chapter 4

k	uniformity coefficient, 1
x_i	rainfall of measurement positions, mm

x	average rainfall of measurement positions, mm
n	number of measurement positions, 1

Chapter 6

α^i	slope angle of slice, (°)
a	acceleration, m/s²
c'	effective soil cohesion at the potential slip surface, kPa
$c(v)$	velocity-cohesion module, N
η	viscosity, Pa · s
ϕ'	effective friction angle at the potential slip surface, (°)
F	slip force, kN
H_i	height of slice, m
h_i	height of the ground water table of slice, m
m	mass of shear zone, kg
n	bulk porosity of the landslide mass, 1
n'	total number of observations, 1
P_{wi}	pore water pressure at the base of slice, kPa
r_d	unit weight of dry landslide mass, kN/m³
r_i	unit density of landslide mass, kN/m³
r_W	unit weight of water, kN/m³
T'	resistance force of the whole slope, kN
$v(t)$	mean velocity at time t, m/s
v	velocity, m/s
w	width of slice, m
Y_i^{obs}	ith observation for the constituent being evaluated, 1
Y_i^{mean}	ith mean of observed data for the constituent being evaluated, 1
z	height of shear zone, m